GW01157376

PACKAGING FOR ELECTRONIC PRODUCTS

DESIGN MEDIA PUBLISHING LIMITED

CONTENTS

PREFACE

Create a Memorable Package Experience to Attract Consumers

Package design is a unique form of art. It is the first opportunity to communicate with your targeted group of consumers and to make an impact on people who haven't had a chance to actually see the product. Therefore, enticing a potential consumer to desire the product inside becomes the primary design goal. The package has to stop people in their tracks and make them want to further explore the goods.

Many companies fail to understand that their product's package is the first impression that decides the overall product experience. It needs to make consumers want to open the box by creating instant anticipation - by the time they reach the product, they already feel excited and want to see and learn more.

Which elements are crucial for a successful package?

For every company, branding is the key to success, and product packaging is an effective tool to visually translate the core values of the product and the brand. Eye-catching designs and images are vital at the most basic level because they deliver the message in a direct way that affects consumers' decision-making processes.
Equally important is a solid and carefully devised structure that allows for easy access to the products while providing sufficient protection during the shipping, storage, and in-store display processes, especially for high-technology products, which need special protection.

This is also relevant to the product's in-store life cycle. People have to believe that the product is trust-worthy to make a purchase, and a firm, well-constructed box reinforces this feeling.

Green Design

A new generation of consumers is concerned about the ecosystem. It is our responsibility to design a package that leaves the smallest carbon footprint possible while still maintaining its core attributes. So our next question is this: how do we reduce carbon footprint?

This can be done from three perspectives: choose eco-friendly and recyclable materials; save more energy and reduce waste by using fewer components and simplifying the manufacturing process; and use smaller boxes. Think carefully about how to facilitate the shipping process as well as how to reduce unnecessary space inside the box for less fuel oil consumption and pollution.

What is the difference between designing a package for technical products and for other product genres?

The principles for package design are actually the same across all product categories - to create a memorable out-of-box experience and enhance users' appreciation for the products. You can develop additional approaches to meet a range of needs.

For example, if the boxes are mainly for shipping and storage, then the design strategy will focus more on protection and size reduction, while packages for in-store display will involve more consideration of the package's appeal for potential consumers.

There are three crucial questions to contemplate before designing a package for a technical product: 1) why do consumers need this product? 2) Will the product work for them? 3) What are the specifications to justify the cost?

When it comes to technical products, too many companies try to include all the technical details on the front of the box, but this "overkill approach" dilutes the aesthetic and branding experience, as well as providing unnecessary technical information where it doesn't belong.

The top priority should be creating an immediate "I want it" response. Many people may be initially intimidated by new technologies. Thus, it is the designer's mission to come up with an attractive and approachable box front image and design and to reduce technical information to a minimum.

Another major goal is to simplify the complexity of information as much as possible. Don't overwhelm consumers with too many messages. The following rules of thumb ensure that the right messages appear on the right planes of the box: core values and major product features should be placed on the front and the top planes, the secondary features should appear on the sides, the simplified technical specifications should be placed on the back, while the regulatory information should appear on the bottom plane.

Prototyping! Testing! Iterations!

Keep in mind that the unboxing process also creates a powerful first impression of the overall product experience. In order to achieve perfection, there are no shortcuts – you will need to go through a detailed process of research, creating prototypes, testing your design, and iteration. This is precisely the process that CRE8 DESIGN uses every time we design a successful new package for our clients – one that is attractive, ergonomic, and intuitively functional.

Kris Verstockt
Founder& Executive Director at CRE8 DESIGN

Kris Verstockt is the founder and senior tactician at CRE8 DESIGN. Bolstered by over 20 years of design and project management experience, the CRE8 portfolio includes successful partnerships with Best Buy, DELL, Nokia, Corsair, Acer, ASUS, GBC, Primax, and many others. Both Kris and his firm have won numerous awards and accolades (iF, Red Dot, IDEA, G-mark). Kris speaks at conferences, leads seminars, and has been invited on various occasions to judge renowned programmes such as the iF awards. He has also lectured at the distinguished Shih-Chien University. Kris Verstockt studied design in Antwerp and London. In 1994, he took a product design position with Primax, a 10,000-person Taiwanese electronics firm where he was in charge of building up an industrial design team. He founded CRE8 DESIGN in 2001.

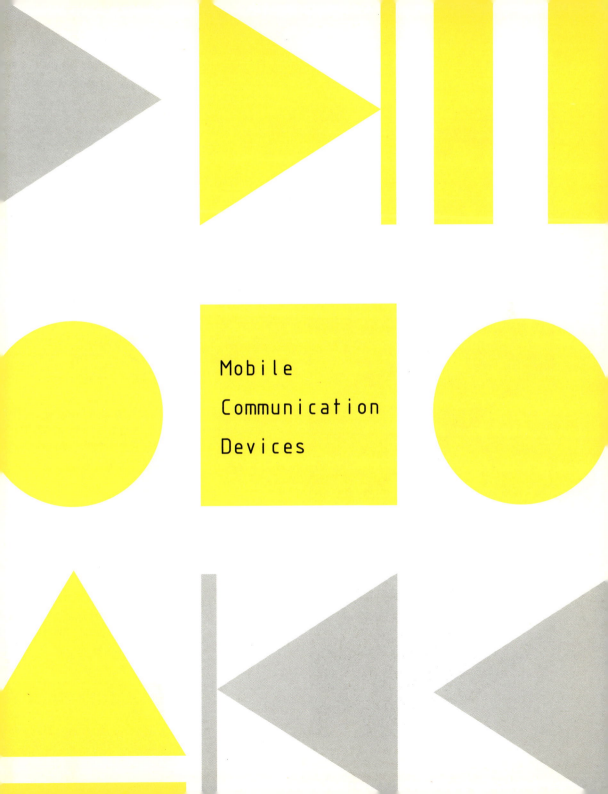

Mobile

Communication

Devices

Out-of-box Experience

Orange asked Make it Clear to design a complete out-of-box experience for technology that made paying with a mobile phone a reality for the first time. Make it Clear created packaging and collateral featuring set-up guides, demo Quick Tap tags and a straightforward process based on a strategic, linear path. Helping consumers to understand the relevance of the new technology even before they had opened the box, the experience made sure key functionality was adopted immediately. The set-up process was made intuitive taking consumers on a simple and quick user-journey.

Design Agency: Make it Clear. Designer: Sarah Edwards. Client: Orange. Nationality: UK.

Sprint EVO 3D

The major graphics on outer white box are elegantly embossed. The inner box has colour fractal graphics that reference the first ever colour 3D LCD screen. The sustainable packaging attributes: packaging is 100% recyclable, with 37% post-consumer recycled paper and printed with soy-based inks.

Design Agency: Deutsch Design Works. Designer: Eric Pino. Client: Sprint. Nationality: USA.

Sprint White Core

DDW's strategic packaging redesign implemented new standards for improved recyclability of 30 million phone boxes per year including: reduced package size and ink coverage, removed plastic laminates and plastic trays. To tell the Sprint brand story, DDW developed the "real life" series of packaging that displays visual storytelling about the intersection of humanity and technology.

Design Agency: Deutsch Design Works. Designer: Eric Pino. Client: Sprint. Nationality: USA.

Sprint Accessories

Sprint has been engaged in embracing sustainable practices and green packaging. DDW's strategic packaging redesign implemented new standards with great annual green savings, especially cost savings.

Design Agency: Deutsch Design Works. Designer: Eric Pino. Client: Sprint. Nationality: USA.

Sprint Epic

Packaging design for a new product of Sprint: Samsung Epic 4G
cellphone. The panoramic design displays the device across three
sides of the box. An innovative design approach that would draw
customers' eyes and present the cellphone to them in an interesting
way.

Design Agency: Deutsch Design Works. Designer: Pauline Au. Client:
Sprint. Nationality: USA.

Sprint Green Phone Family

The design palette uses a family of eco-friendly graphic elements that vary across the four devices in this series. The sustainable packaging attributes: packaging is 100% recyclable, with 70% post-consumer recycled paper and printed with soy-based inks. These phones are made from recyclable materials, with 20% to 30% of the outer casing made of post-consumer recycled plastic.

Design Agency: Deutsch Design Works. Designer: Eric Pino, Pauline Au, Erika Krieger. Client: Sprint. Nationality: USA.

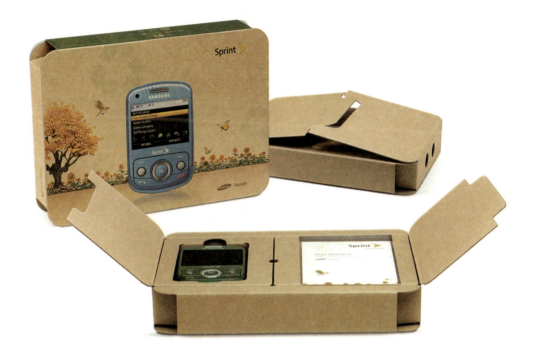

Sprint Reclaim

The Reclaim package complements the more sustainable phone (made from 80% recyclable materials) and makes the total package even friendlier. The outer packaging and the phone tray inside the box are made from 70% recycled materials. All printing on the box as well as the warranty information are printed with soy-based ink.

Design Agency: Deutsch Design Works. Designer: Pauline Au. Client: Sprint. Nationality: USA.

Way and Star Dual TV
Cellphone Packaging

The cellphone Way and Star Dual TV was created to meet the need of a most popular social class. With a price more affordable for the lower classes, there was a need to create a more striking and attractive packaging for classes C and D. From there, the concept of prominence was used, where the colours call attention quickly enough on the shelves.

Design Agency: Multilaser. Designer: Rafael Guímaro. Client: Multilaser. Nationality: Brazil.

Mercury Cellphone Packaging

From a study of competition in the market of mobile phones, was created a new package for the mobile phone Mercury, the first Android phone from Multilaser. The main point adopted for the creation of the packaging was a modern communication through a clean package, with simple elements and objectives, always focusing the attention to the image of the mobile phone.

Design Agency: Multilaser. Designer: Rafael Guímaro. Client: Multilaser. Nationality: Brazil.

MEU Celulares Package Identity

Development of two lines of visual identity for packaging of "MEU Celulares" cellphones. Bright colours dominate the packages.

Design Agency: IP I CONTEÚDO. Designer: Thiagodsmax Client: MEU Celulares. Nationality: Brazil.

Pantech Smart Phone Package Design

This smart phone package design was made during the internship at Pantech. Companies need to suggest new living value to customers as consumers desire for various drive, lifestyle, and purchase behaviour. Often the mobile phone functionality, price, and specifications are equivalent from one to another; therefore, it is important to have a premium value with an element of design. This package design was targeted towards young generation who consider their own unique individuality and design as the most important element; therefore, the designer tried to avoid the feeling of black and white, and minimalistic technology that can be easily visualised when people think of smart phones. The use of primary colour allows the consumer to recognise its form and colour even from a long distance. The pictogram makes new customers feel intimacy to the brand, Pantech. Through designing these sets of package design the designer tried to express the feeling of toy planet with its shape and colour. The designer wanted young generation customers to enjoy as they open the package with fun.

Design Agency: Pantech. Designer: Sangyee Lee. Client: Pantech. Nationality: South Korea.

Z320i 55DSL Limited Edition Cellphone Packaging

In a hi-tech collaboration with Sony Ericsson, 55DSL has launched the Z320i 55DSL limited edition cellphone. The clamshell handset takes its style cue from the aesthetic of 55DSL, the street-wear sibling brand of the Diesel Group. The packaging includes a personalised cover and lanyard.

Design Agency: Quasidesigner.com. Designer: Luca Mathia Bertoncello. Client: Diesel S.p.A. Nationality: Italy.

Nokia i1

Nokia i1 is a personal project to develop the design and packaging of the phone. To date, the design of devices has become boring and no longer causes a "mirror" of emotions. This version has a peculiar shape with modern technical specifications. The designer believes that packaging should reflect that the image of the goods is not less important than the product itself. So pack Nokia i1 has the following format. This project brings us back to an earlier state in which we first got in contact with the models of Nokia. Mirror, dynamics and individuality. The design should be one step ahead of you. Let it guide you to perfection.

Designer: Stas Bordukov. Client: Nokia. Nationality: Russia.

eyo Cellphone Packaging

A custom cellphone package. Four models are available and the customer can choose which box he wants. The idea was to bring to the package the fun and the customisable idea that the EYO FUN® cellphone brings to his users. Inside the box, the cellphone uses personalised themes and patterns inspired by the box designs.

Design Agency: rco design. Designer: Ricardo Colombo. Client: eyo. Nationality: Brazil.

Optimus Magma Packaging

Optimus changed its corporate identity. The Optimus boomerang symbol was replaced by a new one, called magma, which would be the personification of the brand and its services. Optimus was not only interested in knowing people's needs and desires, but also committed to deliver a good service. Optimus wanted the consumer to be the centre of all attentions. This cellphone packaging was conceived to be part of the Optimus rebrand. The magma at the centre symbolised the consumer, which was protected by the package. Therefore, the packaging wanted to communicate to the consumer he could be sure that he would be protected and well cared by Optimus. At the same time, the magma was overflowing because it wanted to communicate that he needed to get out of the box, to be different and innovative. The magma 3D surfaces were modelled to permit the package to be stackable.

Design Agency: Euro RSCG Design & Arquitectura. Designer: Ricardo Capote, Carlos Alves. Client: Optimus. Nationality: Portugal.

Optimus Music Packaging

The growing functionality of mobile phones and services associated with music and the importance of music in people's everyday life guided the development of this new package, whose goal is to ensure better identification of the music category in the point of sale. The shape of the packaging – which reproduces the shape of a column of sound – includes a musical message itself. The goal is to create an icon that summarises the spirit that carries the product and simultaneously convey a playful and symbolic imagery to the consumer. Since this package is made of injection-moulded plastic, it is important to design a solution that gives the possibility to reuse it in different circumstances. The design was made with the intention to have an alternative use that extends the life of this package, making it more environmentally friendly.

Design Agency: Euro RSCG Design & Arquitectura. Designer: Carlos Alves, Cátia Mateus. Client: Optimus. Nationality: Portugal.

Vodafone Packaging

Vodafone is the world's largest mobile communications brand, with over 330 million customers across 21 countries. As part of a wider global brand identity refresh, The Partners created a packaging design solution. Each pack design focuses on the products' "unique selling point" and uses creative and emotional metaphors to promote the product and its benefits; busy bees, sociable shoals and protective feathers. This results in each product having its own recognisable identity and creates a more engaging experience with the customer.

Design Agency: The Partners. Creative Director: Greg Quinton. Designer: Kevin Lan, Freya Defoe, Sophie Hayes, Jessica Harvey, Tom Leach, John Molesworth, Dominic Davidson-Merritt. Client: Vodafone. Nationality: UK.

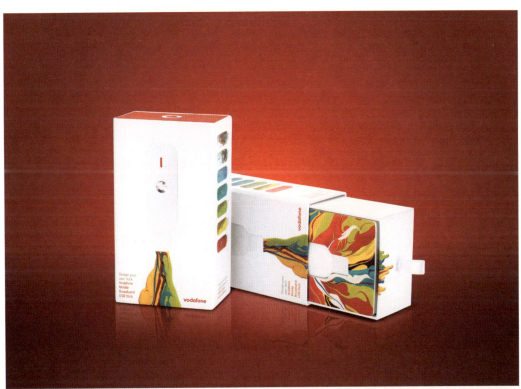

1

2

5

6

Magic Cube

This piece was a specific requirement of the client, in which he wanted from a conventional bucket to give a variety of ways to display.

Design Agency: JWT Bogotá. Creative Director: Javier Hincapie. Designer: Miguel Angel Pérez Murcia. Client: Nokia. Nationality: Colombia.

3

4

7

8

Securo Video Door Phone Packaging

Octocon Systems approached Creative Caffeine for the branding and designing of packing for their new product Sesuro Video Door Phone. The designers believe that when choosing one product over another, the design of the packaging influences the customer's decision far more than what they realise. When creating the package, the designers tried to ensure that they are conveying the necessary information about the contents and quality of the product.

Design Agency: Creative Caffeine. Creative Director: Sooraj Krishnan. Designer: Ranjith Alingal, Jishnu Krishnan. Photographer: Sindhur Reddy Client: Otocon Systems Pvt. Ltd. Nationality: India .

Sony Ericsson Packaging

Packaging and booklet design for Sony Ericsson. The pocket-sized compact Xperia X10 mini and mini pro Android smart phones have been designed for one hand navigation with quick customisation and access to top apps and entertainment.

Design Agency: Burgopak. Creative Director: Dane Whitehurst. Designer: Joseph Malia, Dane Whitehurst. Client: Sony Ericsson. Nationality: UK.

Droid Packaging I

Over the past few years the Motorola Droid has become one of the most technologically advanced brands on the market. In order to reflect the technological branding of Droid in a more impactful way, the designer covered areas of the package in aluminium, and foil stamped the logo. The sturdy composite frame of the package also increases the product's perceived value, heightening the consumer's overall Droid experience.

Design Agency: Jon Marquez Graphic Design. Nationality: USA.

Droid Packaging II

Packaging design for Droid Eris smart phone and limited edition R2-D2 Star Wars Droid smart phone, both by Google and Verizon. The folding diagram clearly depicts the structure of the package.

Design Agency: McgarryBowen. Creative Director: Michael Cannova. Designer: Jason Borzouyeh. Client: Verizon. Nationality: USA.

OGO 2.0

Priced under S90 or bundled with a pre-paid service, the OGO 2.0 is an affordable, mobile messaging device targeted at the teen generation who prefer texting to talking. Using a Bluetooth headset, the OGO is also a mobile phone, yet the audio features are subdued intentionally for a clear differentiation from mobile phones – a critical marketing requirement made by the mobile carriers. The packaging design features a clear black-and-white colour combination, matching the popular simplistic aesthetics of the young generation.

Design Agency: NewDealDesign. Creative Director: Gadi Amit. Designer: Laura Bucholtz, Yoshi Hoshino. Client: IXI Mobile. Nationality: USA.

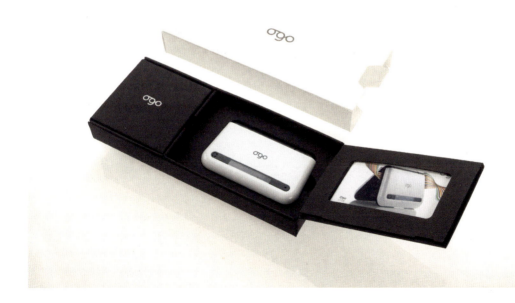

Giftbox KANSO M300 Outdoor Cellphone

KANSO M300: an outdoor cellphone produced to endure rough, moist and dusty environments. The phone is ideal for outdoor activity such as fishing, hunting or trekking. The giftbox design appeals to consumers in that field of interest.

Design Agency: sanseliv. Designer: Charlotte Holst Andersen. Client: SOKAN telecom. Nationality: Denmark.

Telenor Mobile Broadband

The package works as a grab-and-go product in the shop. It contains cash card, modem and everything else you need for surfing with Mobile.

Design Agency: Garbergs. Creative Director: Petter Ödeen. Designer: Jonas Bäckman, Johan Wilde. Client: Telenor. Nationality: Sweden.

uBear

The design brief for the uBear logo was to create an identity that would be both adaptable and instantly recognisable. The logo consists of a stand-alone Bear mark hugging the letter U, as well as a custom sans serif logotype. A unique, bold and bright-colour palette was used throughout the packaging and branding to help distinguish it from similar products on the market. The use of gloss varnishes and foils across the packaging helps create interest on the shelf. In addition to the identity and packaging, a fully responsive website was designed and built to allow a seamless user experience.

Design Agency: Hype Type Studio. Creative Director: Paul Hutchison. Designer: Paul Hutchison, Mark Bloom. Photographer: György Körössy. Client: uBear. Nationality: USA.

Computers and
Accessories

Hanvon Europe Packaging

Hanvon, leading manufacturer of graphics tablets and numerous other devices, entered the European market and asked Twintip to redesign its identity and packaging to make it coherent with European taste and showcase its products in an excellent manner. For the product packaging the inspiration was found in hand drawing creating a graphic style able to position Hanvon's product at a higher level, according to the sophisticated European taste. The brand image was then translated into digital for the development of the e-commerce website. Contents are written using the technique of article marketing, with special attention to SEO.

Design Agency: Twintip. Creative Director: Stefano Guerrieri Designer: Stefano Guerrieri. Client: Hanvon Technology Co., Ltd. Nationality: Italy.

X-Doria Packaging

X-Doria is an international electronic accessories brand. They came to Knoed Creative in need of a complete brand overhaul including a new vision for the brand, redesigning the logo, designing a new packaging system and a shiny new website. With competitors like Incase and Griffin, Knoed had to make sure their packaging turned heads without alienating them from the market. Careful consideration was also given to the logo, making sure it would work well across all applications like embossing, stamping and embroidery.

Design Agency: Knoed Creative. Creative Director: Kim Knoll, Kyle Eertmoed. Designer: Kim Knoll, Kyle Eertmoed. Client: X-Doria. Nationality: USA.

Veiocity Packaging

The packaging design is for a laptop computer. Medley geometric figures are stacked. The designer intended to use these graphics to reflect the feeling of modern science and technology. The green sign is put on the package, attracting attention from customers at first glance.

Designer: Agnes Veles. Client: Askul Corporation. Nationality: Sweden.

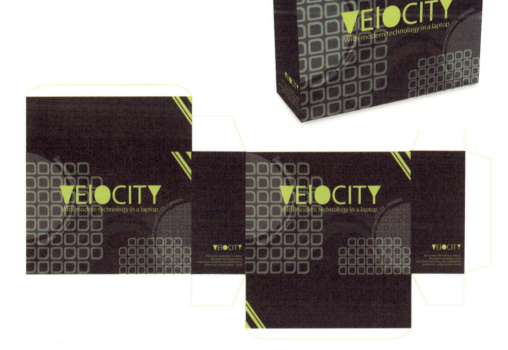

MacBook Pro & iPad Cardboard Case

Recyclable gift for anniversary of Tomas Bata University in Zlín. The case is made from one piece of cardboard and antiskid polypropylene foam tape. It is environmentally friendly and economically inexpensive. This case is dedicated to carry your iPad or MacBook for short distances without any accessories when you don't want or need to take everything with you. It is also able to be used as lapdesk or pad. Antiskid tape with logo of anniversary will hold your Macbook or iPad safely in various angles.

Designer: Michal Marko. Client: Tomas Bata University. Nationality: Slovakia.

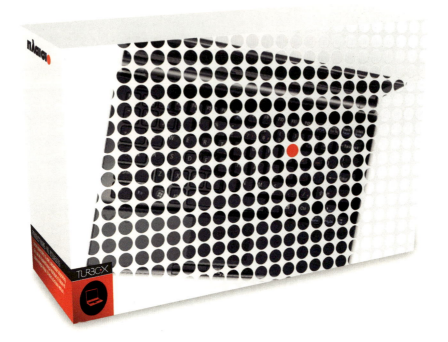

Turbo-X Packaging

Turbo-X, property of Plaisio S.A., is among the top selling consumer electronic brands in Greece. For this project, the designer has developed the complete redesign of the brand visual identity and top products (such as desktop PCs, laptops, monitors, keyboards, mice, and TVs). Two product family concepts have been developed: Bloom (white) and Performance (black), both for the high-end European market. The design phase involved an innovative tactile concept which added subtle and sensorial details to every family product.

Design Agency: Sublimio - Unique Design Formula. Creative Director: Matteo Modica Designer: Matteo Modica. Client: Turbo-X. Nationality: Greece.

iHome Wirless Type Pro Laptop Style Keyboard Case

For iHome's line of Bluetooth keyboard cases the designer came up with a slide-out design that also incorporates a door and window. The inner tray is made of corrugated cardboard and houses the product as well as an envelope that contains instructions and a micro USB charging cable. The outer sleeve is made of 300 gsm paper with a clear PET window that allows costumers to view the product without sliding out the inner tray.

Design Agency: Lifeworks Technology. Designer: Tyler Beichner. Client: SDI Technologies. Nationality: USA.

Dell Packaging

Dell was looking for a solution on how to package their laptops for back to school. It would bundle a laptop, mouse, accessories, etc. for college students to purchase on campus. The packaging you see here was the solution. BBDO came up with an icon system for Dell to use across campuses. BBDO used this same approach in the packaging. It was used as a pattern or wall paper but arranged in a flowing and organic fashion as almost a sea of clouds or thoughts. BBDO also used the laptop on the outside of the box and designed it so when lined up the boxes the laptops would match up and look like one laptop.

Design Agency: BBDO Atlanta. Creative Director: Paul Huggett Designer: Jeff Oehmen. Client: Dell. Nationality: USA.

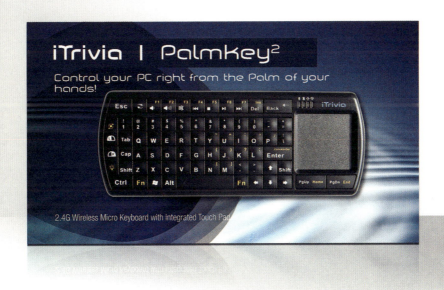

iTrivia Palmkey Packaging

The idea behind this concept was to focus the new branding of this product on the market. Therefore, the designer decided to adopt a new "fresh" colouring designing in combination with the oval shape of the product. The box is already in production with no content, so designing a sleeve around the box was the best idea.

Design Agency: Dian Creative Design. Creative Director: Willy Wong. Designer: Willy Wong. Client: Xitac. Nationality: The Netherlands.

ARC Mouse Packaging

. . . .

Positioned as a lifestyle product and catered to a young target audience the mouse is featured intangibly in a clear-folded plastic box to convey the premium nature of the product. Unlike any other product in this category, the mouse is placed sideways highlighting the innovative folding mechanism and beautiful profile.

Design Agency: SONIC Design. Creative Director: Ralf Groene. Designer: Klaus Rosburg. Client: Microsoft. Nationality: USA.

Duex Mouse Packaging

. . . .

Duex is an electronics and computer goods company from Brazil.
Ghana Branding developed a new colour palette and graphic identity
for the brand, aiming for a warmer and more dynamic look that could
adapt well to the company's extensive product line.

Design Agency: Ghana Branding. Designer: Giancarlo Meneghini.
Client: Duex. Nationality: Brazil.

Rock Mouse

Energetic and clean fresh waters are wat[...]
touchmouse. you will walk along the river[...]
moment into something extraordinary fro[...]
touchmouse.com

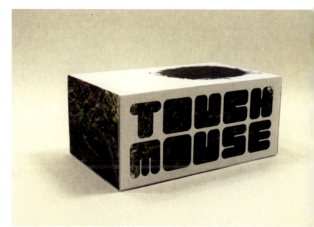

Touchmouse Packaging

Recently, people have lost touch with nature. The reason could be busy lives and the development of electric gadgets. Touchmouse is a brand-produced computer mouse with natural materials. By holding the mouse, it will connect people to nature in their room.

Designer: Rene Kim. Nationality: South Korea.

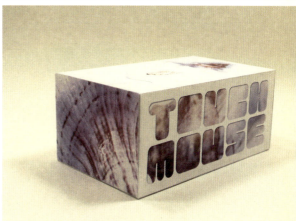

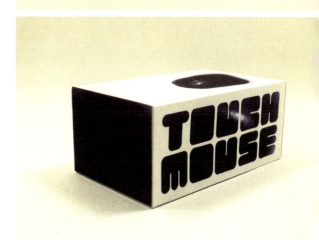

Recycled Mouse Packaging

....

This is a school project. The team of three members did a research on Microsoft electronic products packaging design and noticed that the problem existing in plastic packaging is too hard to open. In addition, some of the parts are wasteful. If using recycled paper or invisible materials, it may couldn't communicate with consumers well. They then decided to design a new line of products, which used sustainable materials and would be more friendly in communication and environment. They designed the new logo type which illustrated the letter "G" as a leaf shape. The front outside piece is sliding and allows people to open it easily. It also has a small holder on the top for saving more plastic bags, to be environmentally friendly.

Designer: Xiaoxi Li, Yun Wang, Emannuel Saka. Client: Microsoft. Nationality: USA.

Starck Optical Mouse Packaging

Swerve were asked to create a new packaging format to position the Optical Mouse as a flagship product. Swerve saw an opportunity to present the post-modern, minimal mouse as a work of art and giftable item. A hybrid type of package incorporating features of both a box as well as clamshell was developed using Fabergé imagery, fused with a photo frame structure. The new design was taken from concept to market within six months and became a treasured collector's piece. Winner of Gold IBPA (International Brand Packaging Awards).

Design Agency: Swerve. Client: Microsoft. Nationality: USA.

Corsair Gaming Mouse Package Design

To create dynamic vibes that appeal to the gaming enthusiasts, CRE8 DESIGN adopted a large diamond-shaped display window to provide an all-round view of product details and a triangular prism box to echo its geometric theme. The blister that holds the mouse is designed to expose the shape of the product, allowing consumers to experience the gripping gestures before unboxing it. The backside of the package incorporates a cut-out window to showcase the key feature of the mouse – the aluminium frame.

Design Agency: **CRE8 DESIGN**. Client: **Corsair®**. Nationality: **USA**.

The Wallee

The idea was to create a simple and charming packaging concept for the Wallee's whole range of iPad Accessories, which illustrates the playful, efficient and easy-to-use character of their products. A thick cardboard was chosen for the different boxes and, finished with screen-printing to reduce the use of plastic materials.

Design Agency: Denise Franke. Client: The Wallee Ltd., Melbourne, Australia. Nationality: USA.

Solo Gateway Packaging

Solo Gateway is a Greek brand providing IT services and products to small, medium and large businesses. The creative concept behind this project takes inspiration from sleek design and professional look. The Solo Gateway Unique Design Formula mixes: Concept Development, Identity Creation, Corporate Image, Product Design, Packaging Design, Online Experience and Branding Applications.

Design Agency: Sublimio - Unique Design Formula. Designer: Matteo Modica. Client: Dot Kite / Solo Gateway. Nationality: Greece.

Vincent Gift – USB Packaging

The USB packaging design is meant to reflect pureness, freshness and naturalness. This is a gift of the annual meeting of Vincent Corporation.

Designer: Agnes Veles. Client: Vincent Corporation. Nationality: Sweden.

How Stuff Works: Podcast Box Sets

....

This rebrand project focuses on communicating How Stuff Works' desire to demystify the world in a clear, unbiased, easy-to-understand format. The design shows the simplicity and range of information that this company offers. Content would be available in "best of" box sets, containing video, audio and other media on a USB, accompanied by a small book of articles and other printed.

Designer: Ilana Addis. Photographer: Douglas Lloyd Photography. Client: How Stuff Works. Nationality: USA.

Highlighter
蛍光ペン
10 pcs
イエロー

Highlighter
蛍光ペン
10 pcs
グリーン

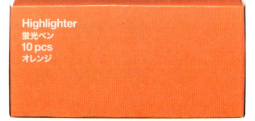

Highlighter
蛍光ペン
10 pcs
ブルー

Highlighter
蛍光ペン
10 pcs
オレンジ

Askul

The problem was that Askul's own brand of products was not highly visible in the product catalogue, which negatively affected sales. With Scandinavian design as a starting point the products' common denominators are function, simplicity and clear communication. The typography has distinctive elements of colour and graphics, which makes the packaging eye-catching within both the extensive catalogue, and the home or office environment.

Design Agency: BVD. Client: Askul Corporation. Nationality: Sweden .

Time Capsule of Barcelona USB Packaging

....

This packaging was designed as a souvenir for the inauguration ceremony of the time capsule of Barcelona, commemorating the 150th anniversary of Cerda's Eixample. The idea was to emulate the real capsule and still being aesthetic after its use. Inside it there was a USB with information relative to the time capsule.

Design Agency: ONIS Design Studio. Designer: Jose Mª Solanes, Jordi Rabal, Pau Jordà, Juan Izquierdo. Client: Barcelona Centre de Disseny, Ajuntament de Barcelona. Nationality: Spain.

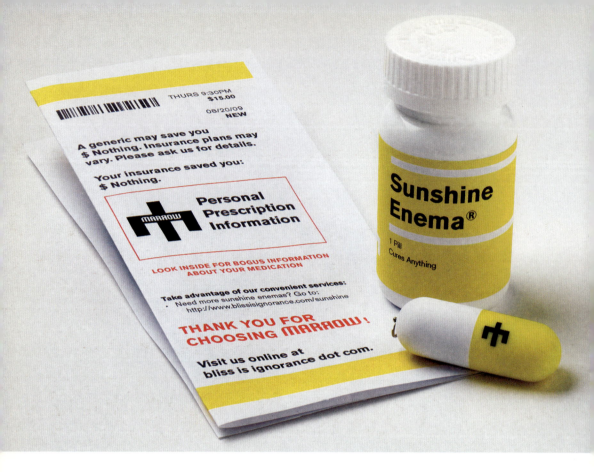

Sunshine Enema (Master Cleanse Kit)

Most people buy their music online – so if you want to sell physical product, CDs just aren't enough anymore. The in-hand experience needs to seduce the buyer. The designers felt that a digital application for the music would not only be the most convenient for the listener, but that the pill-shaped "enema" really brought the album's theme home. It also gave the band a way to add new content including wallpaper and posters Rich Pageant designed for them as well as any new tracks they wanted to offer.

Design Agency: Rich Pageant. Client: Marrow. Nationality: USA.

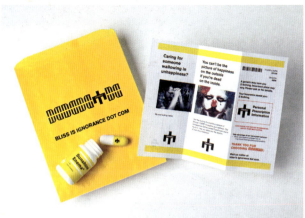

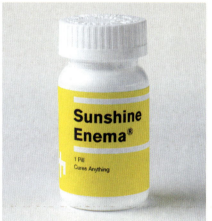

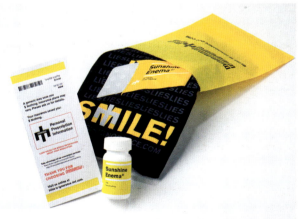

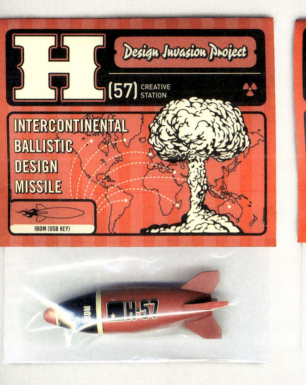

H-57 IBDM USB Key

USB flash drive created as a self-promotional gadget. The rocket-shaped design and the graphic design of the packaging are inspired by American sci-fi movies of the 1950s and 1960s.

Design Agency: H-57 Creative Station srl. Nationality: Italy.

Luxure Media Pack

Branded box created for fashion and lifestyle magazine Luxure. The box opens to reveal belly-banded postcards of photography from the publication and a laser-etched USB stick. The inner tray can be lifted away to reveal the magazine. The box was constructed from Ebony Colourplan over board with a buckram emboss and black foil blocked logo.

Design Agency: Projectarthur. Designer: Mark Hind. Client: Luxure. Nationality: UK.

Fujifilm Performance Series
SDHC Memory Cards

· ·

The packaging for the Fujifilm Performance Series SDHC memory cards was designed to convey the utilitarian use of these cards that were ideal for fast action photography and HD movie. While the product is a great value at an affordable consumer-friendly price, the packaging conveys an appeal to all levels of photographers and photo enthusiasts.

Design Agency: cinquino+co. Creative Director: Ania J. Murray. Designer: Saunak Shah. Client: Fujifilm USA. Nationality: USA.

Cameras

Shoot Me If You Can

....

Fujifilm called iLK to do a limited edition (100) between them. A creative packaging design that the young 18-30 year olds can carry with confidence.

Design Agency: iLK Design. Client: Fujifilm. Nationality: France

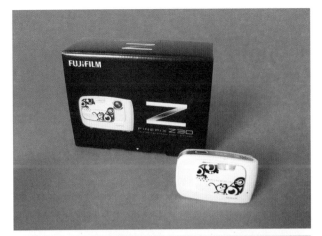

Fujifilm Z30 Digital Camera

Packaging and pattern design for Fujifilm's Z30. Tapping on the trends of the time, this swirly pattern was soon a hit! While the packaging is simple and effective, it compliments the swirly goodness inside. It is part of the latest styles that reach beyond the pallet of clothes and onto their latest and greatest accessories.

Design Agency: cinquino+co. Creative Director: Paul Cirigliano. Designer: Saunak Shah. Client: Fujifilm USA. Nationality: USA.

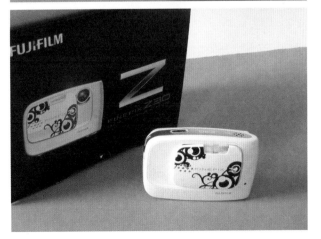

Tracy 3D Camera
EOD-60 AF-S DX NIKOR 18-105mm f/3.5-5.6G ED VR

Tracy Camera Packaging

Colour for the camera is very important. So the designer chose such rich colours as a design element. The packaging gives customers the impression of being very happy and they would like to get this camera. The designer hopes that people can use this camera to shoot brilliant photos.

Design Agency: Fancy. Creative Director: Jesper Green. Designer: Jesper Green. Client: Tracy Camera. Nationality: Denmark.

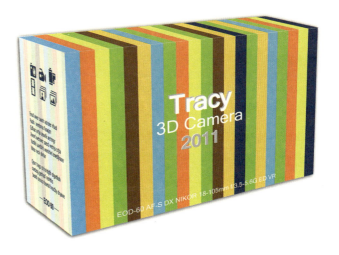

Kodak Conceptual Packaging

As a photographer and person who collects old cameras, it was sad moment when the designer found out that Kodak filed for bankruptcy in January, 2012. He thought about the history of Kodak, the mission of George Eastman, and the modern world today. He decided to conceptualise what would help the brand and be fun, yet important products for photography lovers of all ages. This conceptual packaging is based on Kodak and its history. This brand is known for its historic cameras and making photography available to everyone, not just professionals. To keep the history and mission of Kodak, he chose to mix the look of old cameras with a fun, contemporary design. He chose a monochromatic colour scheme because Kodak is iconic for its yellow gold packaging. The design goal is to attract people and photography lovers of all ages, yet still have the characteristics of Kodak that people already attach it to. The packages are multipurpose and have moveable parts to increase sales and to encourage the consumer to reuse and display the package. The design goal is for the product and packaging to be something older photographers can use and display, yet be fun for children and be a learning device for the history of photography. This creates a higher consumer value for the product. The products he developed are lens cleaning kit and lens case, which also works as a Camera Obscura.

Designer: **Kandace Selnick.** Client: **Kodak.** Nationality: USA.

Diana Mini

The Diana Mini was designed as smaller sister to the Diana+ camera but with some very different features thrown in. The packaging also resembles a small-scale Diana+ packaging but with a box and drawer-style system that pulls out to reveal the camera and a specially created photo book.

Design Agency: Lomographic Society International. Client: Diana. Nationality: Austria.

Diana F+ Tori Amos Edition

The Diana F+ Tori Amos Edition was part of a collaboration between Lomography and the singer Tori Amos. The packaging included a camera, a set of exchangeable lenses, a limited edition CD, a film roll and a photo taken by the musician herself with a Diana camera. The packaging was inspired by the design of a Bösendorfer piano, her preferred instrument on stage.

Design Agency: Lomographic Society International. Client: Diana. Nationality: Austria.

Spinner 360° Motorizer

· ·

The Spinner 360° Motorizer is an accessory for the Lomography Spinner 360° camera that allows the photographer to take panoramic photos remotely using a remote control. An intricate inlay was designed to hold the motorized base and the remote control in place. The long rectangular shape of the packaging also reflects the long format of the photos that the camera produces.

Design Agency: Lomographic Society International. Client: Diana. Nationality: Austria.

Fisheye 2

The Fisheye 2 camera was a successor to the original Lomography Fisheye camera with some newly added features. The packaging opens up to reveal the camera behind a plastic bubble. The plastic bubble not only allows you to see the camera inside but is also a reference to the fisheye-distorted photos that the camera produces.

Design Agency: Lomographic Society International. Client: Diana. Nationality: Austria.

La Sardina Czar Edition

The La Sardina is a Lomography camera whose design resembles the classic shape and form of a sardine can. The packaging design was also carefully selected to continue this theme; wooden boxes that are reminiscent of small food crates. The boxes are fitted with a carefully designed inlay that holds the camera and book in place. The bottom wooden panel of the box slides out to reveal the packaging contents.

Design Agency: Lomographic Society International. Nationality: Austria.

Lomo LC-A+ Silver Lake Edition

The Lomo LC-A+ is the Lomography camera that kickstarted the entire movement and the Lomo LC-A+ Silver Lake is a special limited edition adorned with genuine brown leather. The packaging consists of a wooden collector's box with a unique inlay, decorated to celebrate the heritage behind the camera. The package also contains a hardback book, certificate, a manual and film.

Design Agency: Lomographic Society International. Nationality: Austria.

Vintage iPod nano Box

This is a personal project. Everything but not vintage is the new iPod nano touch. The designer wanted to show the benefits of this mini treasure in a vintage box, which ironically shows that inside there's not only music; there is also a camera, but extremelly little.

Creative Director: María Laura Caballero Tejada. Designer: María Laura Caballero Tejada. Nationality: Perú.

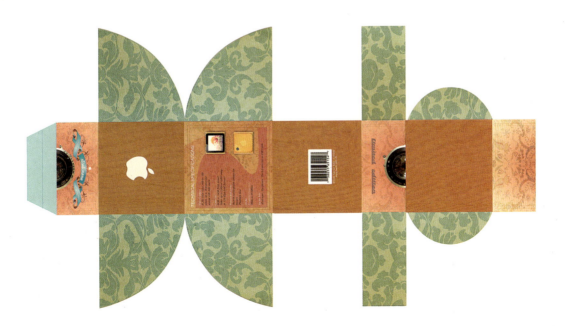

Bonita Camera Packaging

The design brief was to demonstrate the evolving technology advancements within the organisation's key market sector of motion capture. Z3/Studio were keen to use impactful colours within the project mixing gloss and uncoated materials for the box construction. The result was a very tactile and interesting piece of packaging.

Design Agency: Z3/Studio. Designer: Joanne Green. Client: Vicon Motion Systems. Nationality: UK.

Nikon 1 Package & Ads

The "Nikon 1" was built from the ground-up with the advanced features of a DSLR and the compact functionality of a snap-shooter. It became a whole new class of camera, which is mirrored in this launch design with a sense of lightness and bold minimalism.

Design Agency: Thinkdm2. Creative Director: David Annunziato
Designer: Michael Sutherland. Client: Nikon. Nationality: USA.

NO BODY Lens Packaging
· ·

Repackaging of the Olloclip fish-eye lens, as a personalised year-end gift for the studio's best clients.

Message: "NO BODY widens your view. Change angle!"

Creative Bias: As visual communication designers, NO BODY's work consists of creating interest through original point of views. The gift is a wink to invite clients to share their original vision in a friendly quirky way.

Design Agency: NO BODY Visual Communication Design. Designer: Claire Nguyen. Nationality: France.

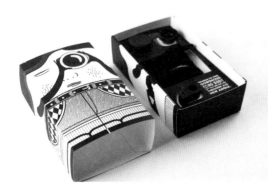

Household
Appliances

Lambretta Veloce

Started in 1847, Lambretta is popularly known as the Italian line of scooters originally manufactured in Milan by Fernando Innocenti. This "Lambretta Veloce" line features a collection of five streamlined appliances that are designed with the same precision and style as the classic Italian scooters. It is named "Veloce", which in Italian, means "speed", referring to the power of these appliances. These sleek and modern additions to your home are not just made to stand out but are built to last, performing with exceptional quality and efficiency. The packaging design mainly uses photography of the products, highlighting the unique style of the products.

Design Agency: Erin Canoy. Creative Director: Erin Canoy. Designer: Erin Canoy. Client: Lambretta Veloce. Nationality: The Philippines.

SWIFT BREWER

EDGE OPENER

- Removable Head
- Bottle Opener
- Knife Sharpener
- 6' Power Cord
- Automatic Shut off
- Anti-topple Base

WWW.LAMBRETTA.COM

MODELLO VL 121

CURVE TOASTER

The Lambretta Curve Toaster delivers an even and even toasting of your slices with its extra-wide and adjustable slots. Easy to clean, this reflective toaster mirrors a Lambretta's orb-cream crisp.

RAPID MIXER

CURVE TOASTER

MODELLO VL105

ZIP CLOCK

The Lambretta Zip Clock is a timeless classic with its dynamic second sweep, polished chrome rims and large black numerals that will keep you working on time.

Caterpillar Home Appliance Packaging

The goal of the project was to extend the identity of Caterpillar into the home of the consumer. Durability, mobility, and commitment are all a part of company's mission as a global provider of machinery and construction products and services. A range of household items were explored throughout the course of the project and brought together using photography and illustration.

Design Agency: Erik Carnes Agency. Creative Director: Erik Carnes, Thomas McNulty. Designer: Erik Carnes. Client: Caterpillar Inc. Nationality: USA.

Hunter Fan Packaging

Research showed that Baby Boomers are loyal Hunter Fan Company customers but Generation Xers are less familiar with this innovator and category leader. Hunter wanted its logo and packaging to pay respect to the company's heritage and premium position in a classic, yet modern, way. It also wanted to differentiate from competitors who had mimicked Hunter's look. The high-energy redesign features an elegant script logo and a "Since 1886" designation. Highly cropped photos "hero" Hunter's high-quality products. The client credits Hunter's dynamic new look to Interbrand's unique design process: "When you marry analytics with outstanding creative, you get a great end product."

Design Agency: Interbrand (Cincinnati office). Designer: Ted Monnin. Client: Hunter Fan. Nationality: USA.

GE Small Appliance Packaging

Rouge 24 found that competitors seemed to focus on male consumers, using tech-heavy language to explain their product features as well as masculine colours like grey and blue (colours that are natural appetite suppressants) on their boxes. While this approach does promote feelings of trust and reliability, it may not inspire hunger to purchase. This series of package design did exactly the opposite.

Design Agency: Rouge 24, Inc.. Designer: Ann Macdonald. Client: Walmart. Nationality: USA.

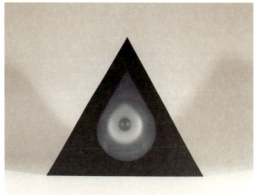

Noir

Noir is a packaging and branding project for a French press coffee maker and two canisters of gourmet beans. The packaging is inspired by the product's formal qualities of transparency and reflection. It creates a fully functional tryptich, appropriate for the spiritual experience provided through the mystical drink.

Design Agency: Analog Design. Creative Director: Brad Breneisen. Designer: Brad Breneisen. Client: Noir. Nationality: USA.

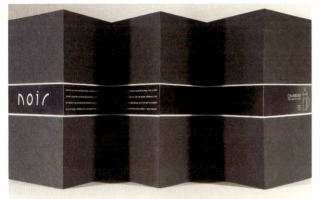

On

A brand from NetOnNet in which the package is designed to function
in its warehouse stores, where the products are never unpacked. All
the sides of the product are therefore shown on the carton – including
the back, as it is important to see what input ports there are. The
colour on the underneath shows the item's product group. The EAN
code, which is incorporated in the logotype, is unique for the product
and can be read at the checkout. The article and model numbers are
also included in the logotype.

Design Agency: Garbergs. Creative Director: Petter Ödeen. Designer:
Jonas Bäckman. Client: NetOnNet. Nationality: Sweden.

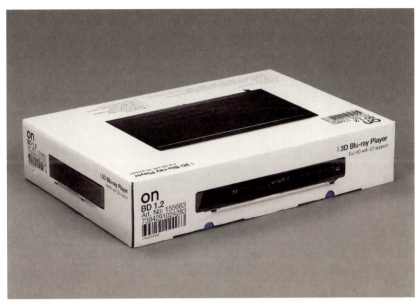

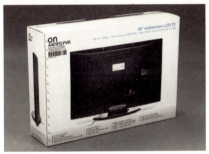

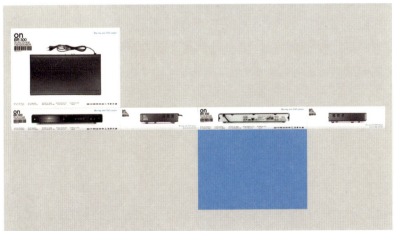

Bell TV Packaging

Packaging that focuses on the concrete advantages that Bell brings each one of its customers every day with its entertaining and diverse offering. People express the complete range of emotions: surprise, enthusiasm, fear, pleasure, wonder, etc. Within the overall goal of minimising the company's environmental footprint, the packaging stands alone; it does not require an additional bag as the different boxes come equipped with a handle.

Design Agency: lg2boutique. Designer: Serge Côté. Client: Rick Seifeddine, Nicolas Poitras. Nationality: Canada.

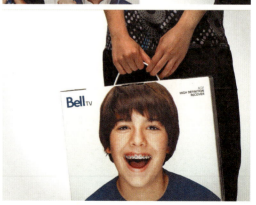

American Red Cross Collection

While working under Etón Corporation, the designer was given the task of designing and marketing the home and road safety product line. This particular collection was part of the American Red Cross and Etón cobranded product line. The main objective was to communicate the product features in a clear and elegant way while still staying consistent with the look of the American Red Cross brand.

Design Agency: R2works. Creative Director: Richie Brumfield. Designer: Richie Brumfield. Client: American Red Cross. Nationality: USA.

Etón Raptor

This product is part of the Etón solar-powered series. The project was issued under Etón Corporation with two main objectives: create a compelling premium brand presence while also communicating the product features in a clear and concise manner. The product itself is a multi-purpose outdoor device that calculates a variety of different weather and terrain figures.

Design Agency: R2works. Designer: Richie Brumfield. Client: Etón Corporation. Nationality: USA.

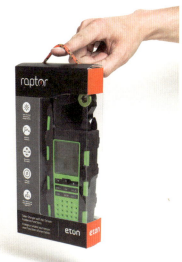

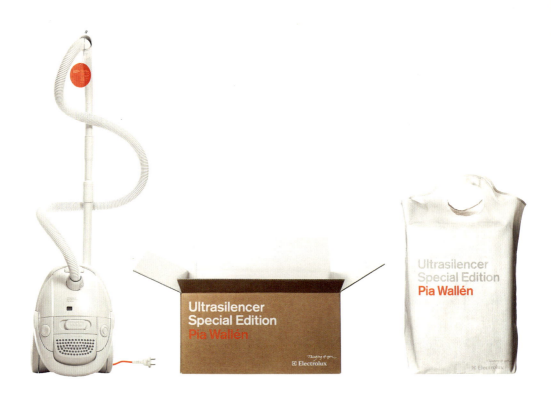

Ultrasilencer Special Edition Pia Wallén

Electrolux needed a graphic identity and packaging solution for the Ultrasilencer Special Edition Pia Wallén. It needed to mirror Pia Wallén's own interpretation of the product, inspired by the quiet sound of falling snow. A typographically based identity was created and applied directly onto the vacuum in a light grey tone. The packaging was turned "inside-out". The interior has a white, glossy surface, encasing the vacuum, and the exterior is brown, natural cardboard. "Special Edition Ultrasilencer Pia Wallén" was screenprinted on the outside of the package in white and orange in order to create an industrial expression in contrast with the white perfectionism of the vacuum.

Design Agency: BVD. Creative Director: Susanna Nygren Barrett. Designer: Carolin Sundquist, Johan Andersson. Client: Electrolux. Nationality: Sweden.

Ultrasilencer
Special Edition
Pia Wallén

...ncer Special Edition Pia Wallén hos
...ckan 19–21.

...folkning av en av världens tystaste
...te ljud hon känner till – ljudet av fallande
...r i signalorange som suddar ut gränsen
...ux.com/specialedition

...my Myllymäki. OSA på party@asplund.org

ASPLUND   ⊟ Electrolux

Gorenje Small Home Appliances Package

Gorenje needed a complete makeover of the packaging for its small home appliances. The designer created a standardised system that would be consistent throughout the range and convey the brand's intrinsic values in a fresh way while offering to the consumer clear, uncluttered information about the product.

Design Agency: Gorenje Design Studio. Creative Director: Blaz Bajzelj. Designer: Blaz Bajzelj. Client: Gorenje. Nationality: Slovenia.

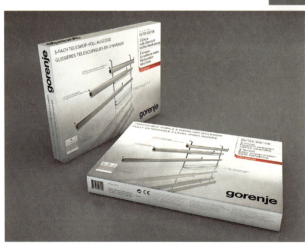

Euro-Pro Kitchen Packaging

The goal of this packaging redesign was to highlight the high-quality stainless steel features of these new appliances. The photography, clean layout design and sophisticated type treatment were used to accompish this. The feel was modern and stylish – products you would want to have on display in your kitchen and not just hide them away in cupboards. Also included in the new packaging was a concentration on "Out of Box Experience" giving the consumer brand touchpoints like a recipe booklet that was an added bonus when you opened up your new cooking appliance.

Design Agency: Euro-Pro In-House Design. Creative Director: Anne Sommers Welch Designer: Anne Sommers Welch. Client: Euro-Pro Kitchen. Nationality: USA.

Ninja Kitchen Packaging

The Ninja Kitchen line of blenders & food processors are one of the most successful products at retail. The packaging really needed to jump off the shelf at retail and make a bold and memorable statement about the brand. The use of industrial pop colours, foil treatment on the logos & down the bevel created a bold brand statement – Ninja is here to stay and we back up our products. Another important feature was the "Out of Box Experience" where Ninja Kitchen is letting the consumer know through beautiful photography and friendly type treatment that Ninja understands their needs, is listening to them and is supporting their healthy lifestyle with easy-to-use recipes, getting started tip booklets and directing them to Ninja website & facebook sites.

Design Agency: Euro-Pro In-House Design. Creative Director: Anne Sommers Welch. Designer: Ryan Lombardi. Client: Ninja Kitchen. Nationality: USA.

Shark Vacuum & Steam Packaging

The Navigator Light and Shark Steam Pocket Mop Light are two recent products from the Shark Floorcare Line. With a high level of performance in a lightweight design, the packaging needed to be sleek while still conveying the quality and features of the product. The predominantly black box helped make the product photography pop, and differentiated the products from competitors on shelf. Bold copy blocks in silver on black conveyed the power and performance of these products. The use of beautifully framed, high-end lifestyle photography speaks to how the styling and design of the product fits seamlessly into even the most modern and refined homes.

Design Agency: Euro-Pro In-House Design. Creative Director: Anne Sommers Welch. Designer: Joshua Hanson, Ryan Lombardi. Client: Shark Vacuum & Steam Products. Nationality: USA.

PowerBall

PowerBall is not just a battery charger but also a handy piece of training equipment for the journey or a possibility to cut stress. Inside, there is an ingenious system. A magnetic mechanism generates elecricity and stores it in built-in accumulator. Hence, there is the opportunity to charge in real time or retrieve the stored electricity. The material sets the mashed-up rubber ball in its native form (circular) automatically. The PowerBall is little, handy and playfully to operate. The design is puristic, modest but also precious and extravagant. Colours of black and white are used to create a clean and clear package, highlighting the image of the product.

Design Agency: Lexon Design Designer: Alexander Schwabenland
Nationality: Germany

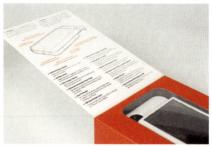

Etón Mobius Packaging

....

This product is part of the Etón solar-powered series. The project
was issued under Etón Corporation with two main objectives: create
a compelling premium brand presence while also communicating the
product features in a clear and concise manner. The product itself is a
solar-powered extended battery-pack and case for the iPhone 4S.

Design Agency: R2works. Designer: Richie Brumfield. Client: Etón
Corporation. Nationality: USA.

Jabra Unity Packaging

Whether it is a new hifi system or a simple charger, you want to see
what it is you are buying. This packaging concept it divided in two:
creating a stage in clear polycarbonate for showing the product and a
backstage for cables and product instructions. The two containers are
fixed by a rubber band and shaped beautifully to embrace each other
creating a strong unity. This packaging concept was a design for a
new universal Jabra charger. However, the project was temporarily put
on hold to focus on a new and more interesting consumer product also
designed by 57N Design.

Design Agency: 57N Design. Creative Director: Anna Michailidis.
Designer: Steffen Mølgaard Larsen. Client: Jabra. Nationality: Denmark.

Bright Thinking

The brief was to design a contemporary light bulb pack promoting the
new energy-efficient LED lighting for the main DIY retailers in the UK
and possibly Europe. The package must fulfil the four main needs of
the consumer; transport, storage, usage and disposal. The response
to this brief was to create a package that would eliminate the use of
plastics within bulb packaging as they are manufactured using non-
renewable sources such as oil and natural gas and also to reflect
the selling point of the LED bulbs which is a better impact on the
environment. The designer then chose to eliminate the various different
individual nets for multi-packs of the bulbs such as packs of two and
four and created an entirely unique design which allows the retailer to
separate the pack using clever perforations to create six variations from
the one net design.

Design Agency: Nick Reid Design. Client: European DIY Retail Stores.
Nationality: UK.

138

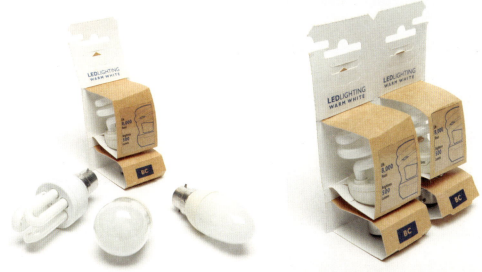

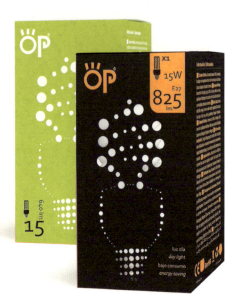

OP Light Bulbs Packaging

OP is the proposed new line of packaging for light bulbs distributed by the Hidalgo's Group and sold by the "Corte Ingles" – the European leader in major department stores. Hache Comunicación focused on making the packaging stand out from the shelves in the big malls by using striking and contrasty colours. Special printing inks and die-cuts were used.

Design Agency: Hache Comunicación. Client: Hidalgo's Group. Nationality: Spain.

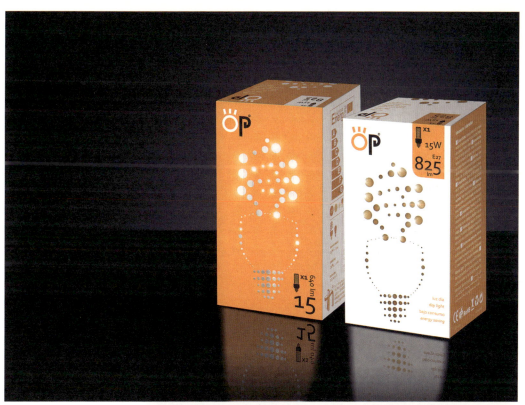

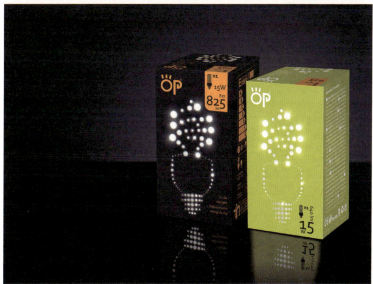

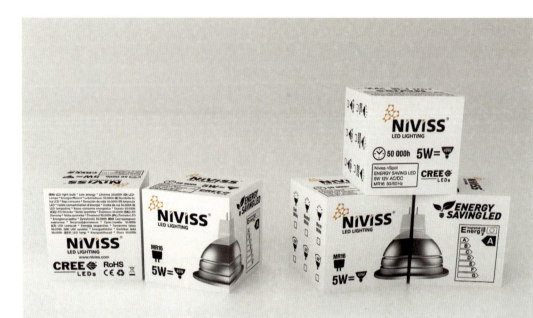

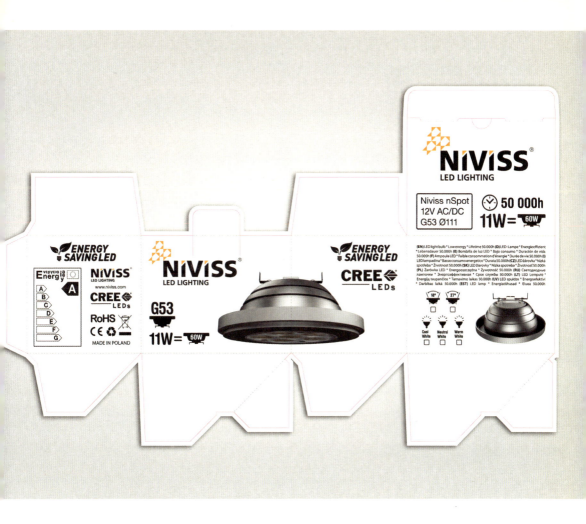

Niviss Spot Packaging
···

Packaging design for LED bulbs. Each version has been prepared in a
different type of size.

Design Agency: RiotHueLab. Designer: Lukasz Szejbut. Client: Niviss.
Nationality: Poland.

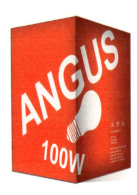

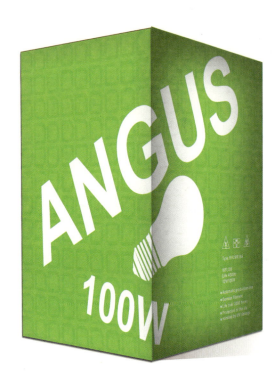

Angus

The design is for Angus bulb packaging, accompanied by bright colours to attract everyone's eye. The packaging design is a combination of the characteristics of the Angus bulbs.

Designer: Simon Webbe. Nationality: UK.

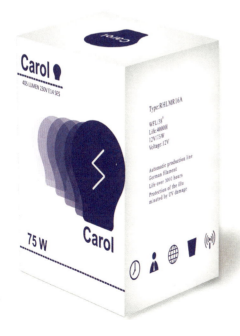

Carol

Design for the Carol packaging.

Design Agency: wonder. Designer: James Blunt. Client: Carol.
Nationality: New Zealand.

Vivid

Protection is a primary function of corrugated paper. It's the first choice for fragile items packaging. White is used coupled with green, so that when people see the packaging design, they'll think of the environmental protection and energy saving. Be green and save electricity. That's the concept for this package design.

Design Agency: wonder. Designer: Kasper Jeffrey. Client: Vivid. Nationality: Denmark.

Veson

The main characteristic of the light bulb is that when people use it at night, it can make the whole room as bright as in daytime. So the designer chose the blue colour for the packaging, matched with white light bulb diagram, to show this concept.

Designer: James Blunt. Nationality: New Zealand.

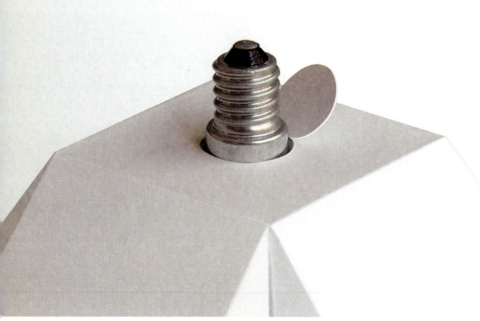

Muji Light Packaging

The design reinterprets the humble lightbulb packaging, aligning it to Muji's utilitarian and inventive approach to product design. Upon opening, the box undertakes an effortless transition into a lampshade, fulfilling a second function. Once the bulb has expired the lampshade makes a third transition back into a box for the bulb's safe disposal.

Design Agency: Ben Cox Design. Client: Muji. Nationality: UK.

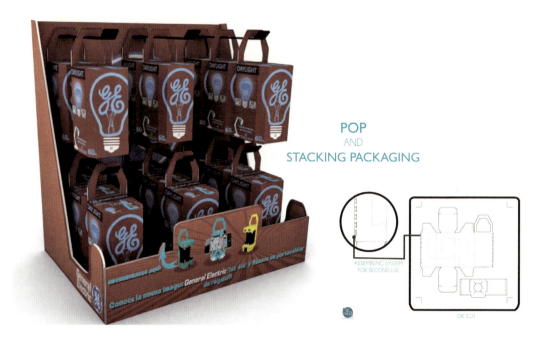

Multi-function Bulb Packaging

This multi-function packaging contains innovative features in form and colour. It is a package for day light bulbs (limited version), and with a special assembly it can become a showy cellphone holder, a practical and entertaining device with three different designs.

Designer: Francisco Lopez. Nationality: Chile.

FIRST USE

SECOND USE

PORTA PHONE

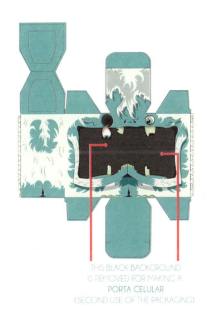

THIS BLACK BACKGROUND
IS REMOVED FOR MAKING A
PORTA CELULAR
(SECOND USE OF THE PACKAGING)

THREE DESIGNS

Blacklight

In 1960s and 1970s, it appeared as part of the psychedelic culture. The designer combined blacklight bulb packaging and blacklight posters to evoke a nostalgic feeling.

Designer: **Rene Kim.** Nationality: **South Korea.**

Solo Plus

Solo Plus is a night time bike puncture repair kit that is contained in a solar-powered light which is charged during the day time. The name Solo comes from the cyclist on his own. The use of the two letter "O"s signifies the solar light torch and the bike puncture repair kit. The positioning of the plus in the logo conveys the added benefit of the two products in the one kit to the solo cyclist and portrays the products' unique selling point.

Designer: Tomas Ashe, Claire Buckley. Nationality: UK.

General Electric CFL Bulbs

Rebranding of Earth Hour. With the support of General Electric's lighting supplies, Earth Hour is creating awareness by informing the consumers to replace their bulbs, save money on their bills as well as participate at the annual event by switching off the lights for one hour to help create awareness for global warming.

Design Agency: **Chulgrafik**. Designer: Chul Lee. Nationality: South Korea.

ČEZ: Children Merchandise

To accompany the huge ATL campaign, ČEZ wanted to create a set of various merchandise targeting children. These are selected materials that were given away or won in competitions. The main prize was the orange lamp itself, brand new ready for its owner to customise it with starter sticker sheet containing eyes and mounts. The lamp came in custom-designed box with the mentioned stickers and Philips bulb. Screenprinted on silver carboard box.

Design Agency: AARGH. Nationality: Czech Republic.

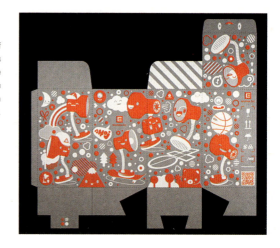

Eco Lamp

Collection of bulbs with new technology: Energy-Saving Lamps. The client find some colours for bulbs with different watts. A hole on lateral is made so that customers can test the light before buying.

Design Agency: jesus-lopez.com. Designer: Jesús López. Client: Noru. Nationality: Spain.

Spectrum Light Bulb Packaging

Spectrum is a fictitious lighting company created for a packing project in the designer's Graphic Design Studio class. The aim of the project was to recreate light bulb packaging to be visually interesting and at the same time protect the bulb. The solution was a strong hexagonal box that housed the bulb suspended in an insert in order to prevent the bulb from touching the walls. A package was created for an LED bulb, a CLF bulb, and a Halogen bulb.

Designer: Nicholas Menghini. Nationality: USA.

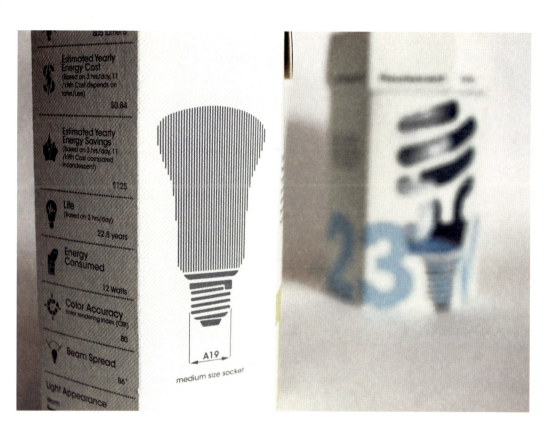

Estimated Yearly
Energy Cost
(Based on 3 hrs/day, 11
¢/kWh Cost depends on
rates/use)

$0.84

Estimated Yearly
Energy Savings
(Based on 3 hrs/day, 11
¢/kWh Cost compared
incandescent)

$125

Life
(Based on 3 hrs/day)

22.8 years

Energy
Consumed

12 Watts

Color Accuracy
color rendering index (CRI)

80

Beam Spread

86°

Light Appearance

A19

medium size socket

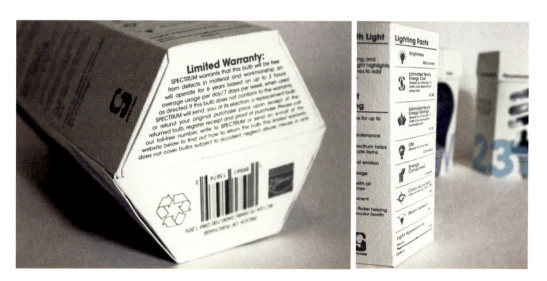

Limited Warranty:
SPECTRUM warrants that this bulb will be free
from defects in material and workmanship an
will operate for 6 years based on up to 3 hours
average usage per day/7 days per week, when used
as directed. If this bulb does not conform to the warranty,
SPECTRUM will send you, at its electron, a replacement bulb
or refund your original purchase price upon receipt of the
returned bulb, register receipt and proof of purchase. Please call
our toll-free number, write to SPECTRUM or send an e-mail at the
website below to find out how to return the bulb. This limited warranty
does not cover bulbs subject to accident, neglect, abuse, misuse or acts

Lighting Facts

Brightness

800 lumens

Estimated Yearly
Energy Cost
(Based on 3 hrs/day, 11
¢/kWh Cost depends on
rates/use)

$0.84

Estimated Yearly
Energy Savings
(Based on 3 hrs/day, 11
¢/kWh Cost compared
incandescent)

$125

Life
(Based on 3 hrs/day)

22.8 years

Energy
Consumed

12 Watts

Color Accuracy
color rendering index (CRI)

80

Beam Spread

86°

Light Appearance

Lucetta Magnetic Bike Lights

Made up of two small magnetic lights, the Lucetta is the new, essential light for your bike. Easy to attach to any bike, the two small lights – one red and one white – switch on with just a click and are guaranteed to stay securely in place even on the bumpiest street. You can select a steady beam, a slow or fast flashing light by simply clicking the light on the bike. When you reach your destination, remove the lights, join them together and slip them in your pocket ready for your next outing. Design by Pizzolorusso. Designs for both its products and packages are simple yet unique, matching the popular simplistic aesthetics of people.

Design Agency: Alvvino. Designer: Alessandro Maffioletti. Photographer: Pierluigi Anselmi. Client: Palomar. Nationality: Italy.

Maglite Flashlight Packaging

Package redesign of Maglite flashlights. A sustainable package using 100% recycled paper tubes. The ink chosen does not include any barium.

Design Agency: Chul Lee Design. Creative Director: Chul Lee. Designer: Chul Lee. Client: Maglite Flashlight. Nationality: USA.

FitBit "Jewel Box" Packaging

··

The packaging for The FitBit is a playful, yet sophisticated approach to graphics and structure. The physical package is a single male/female clear plastic part. Graphics grab the viewer and create a halo of vibrant visuals. The design showcases is constructed from recyclable materials.

Design Agency: NewDealDesign. Creative Director: Gadi Amit. Designer: Yoshi Hoshino, Laura Bucholtz. Photographer: Mark Serr. Client: FitBit Inc. Nationality: USA.

TomTom White Pearl Packaging

Manufacturers of consumer electronics have ignored women as a target group for a long time. TomTom developed the White Pearl specially for them. A luxurious edition and specific content provide an elegant and pleasant navigation. Today Designers contributed significantly to concept, brand identity and packaging design of the White Pearl. The world is your oyster!

Design Agency: Today Designers. Designer: Iwan Kempe. Client: TomTom. Nationality: The Netherlands.

White Pearl
by TomTom

Panasonic Men's Shaver Packaging

Panasonic was a brand with a very segmented Men's Shaver line making it difficult for the consumer to differentiate between offerings and find the one best suited for their needs. Noticing the confusion Panasonic realised the necessity for a unifying image that clearly communicated each product's premium high-tech features. To establish a hierarchy of information the brand name Panasonic was given top billing, while the sharp blade, cleaning and motion features were detailed on side and back panels. A complete packaging restage earned Panasonic placement of their men's rechargeable shaver line with several retail partners – a big win and nice complement to their line of women's shavers.

Design Agency: Group 4. Client: Panasonic. Nationality: USA.

Rea Advanced Cavitation Fat Removal

Rea is an advanced cavitation fat removal product and Matadog's mission was to design the branding and product packaging. The packaging was designed with a vibrant, fresh look by using green and playful artwork that focuses on product benefits.

Design Agency: Matadog Design. Client: Westfold. Nationality: Greece.

Audiovisual Entertainment Products

Tenor Wireless Speaker Packaging

The Tenor Wireless Speaker is a luxury and sophisticated device. It uses wood and Corian to create unique and one-of-a-kind experience. The designer wanted to translate the product language to the packaging and create the holistic experience. In the packaging he used the most iconic element of the product – the speaker grill. It's like a signature of the speaker that defines its uniqueness. So powerful, that there's no need to add anything. It also informs which model of the speaker is inside. Subtle satin finish mimics Corian and gives the packaging luxurious feeling.

Designer: Rafa. Czaniecki. Client: Tenor. Nationality: Poland.

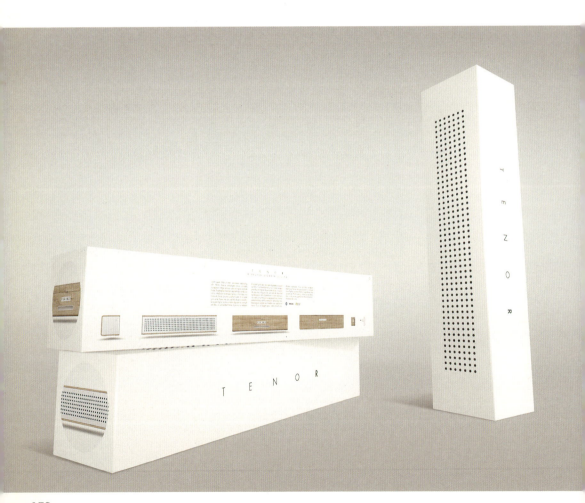

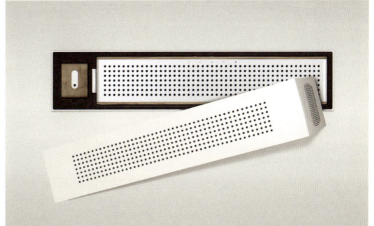

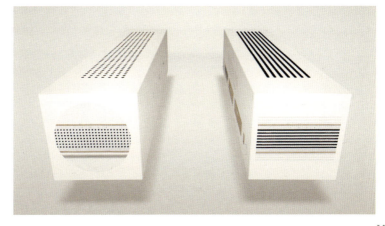

Etón Soulra Packaging

This product is part of the Etón solar-powered series. The project was issued under Etón Corporation with two main objectives: create a compelling premium brand presence while also communicating the product features in a clear and concise manner. The product itself is an outdoor iPod and iPhone sound system that charges via the solar panel.

Design Agency: R2works. Creative Director: Richie Brumfield. Designer: Richie Brumfield. Client: Etón Corporation. Nationality: USA.

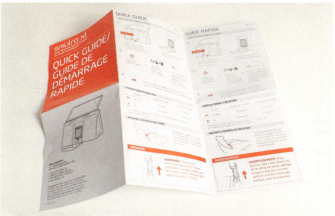

"Einaudi" Series of Packages

It is the design of series of packages for acoustic systems. This project was dedicated to Ludovico Einaudi, the famous Italian pianist and composer. The designer looked for the inspiration for this project in classical music, as it is a symbol of high-quality sound. The main challenge was the creation of various scales of packages. (The sides of the smallest box were 6 centimetres. The largest box was 1200 centimetres high.) The design was based on black-and-white drawings of the lines, which resembled a bar-code and a score (published music) simultaneously. The bright yellow stickers were a striking accent in the project. They are not only aesthetic but also functional object. They had all necessary information about hardware, and as a result it had made the sorting and searching of products in warehouses easier.

Design Agency: Moscow State Academy of Industrial and Applied Art SG Stroganov. Designer: Tatiana Rogatyuk. Nationality: Russia.

Sound!

Long collection of speakers for multiple appliances. The same design for the packages is aimed to be associated with the products family. Blue lines and black colours were used to give a superior sensation of quality.

Design Agency: jesus-lopez.com. Creative Director: Jesús López. Designer: Jesús López. Client: Noru. Nationality: Spain.

Spack Sound System Packaging
with Latin Style

Spack is the first active and portable speaker developed in Chile. It is
a graphic support able to radiate strong sound and visual messages
in its environment. In its first edition, three Chilean artists and one
Colombian are dressing the cardboard with exclusive designs for the
first family of Spack.

Design Agency: GrupoVibra SpA. Creative Director: Benedicto Lopez.
Designer: GrupoVibra SpA Design Team. Client: Design Stores.
Nationality: Chile.

Boosted Tin Can Speaker Project Vol.1

The Tin Can Speaker Project is a functional product intended to offer a platform for artists to showcase their artwork in a unique way. The design of the desktop speaker set represents what may have been your very first experience with communication – just two tin cans and a string. The symbolism in the design relates to coast-to-coast communication and a grassroots interest in art and music. The artists featured in Vol. 1 respectively represent the East Coast with MINT and SERF and the West Coast with MAINFRAME. The fully functioning desktop speaker set is compatible with Boost Mobile MP3 Player phones, iPods, and more. Box packaging, look book, and identity developed for the project focuses on highlighting the customised artwork all while telling the story.

Design Agency: Bay Cities Container. Designer: Nick Holt. Client: Boosted, Boost Mobile USA – Sprint. Nationality: USA.

Loud Mouth

With packaging that reflects a time of families huddled around radios and jukeboxes, the Loud Mouth is a fun, modern-retro solution to a common problem amongst music lovers. Working with your existing headphones and MP3 device, it protects and organises them, as well as functions as a portable speaker.

Designer: Mary Boyle, Connie Shim, Christina Xu. Nationality: USA.

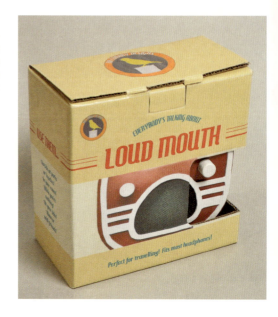

JOY Bluetooth Speaker Package Design

....

Packaging an elite class of digital products that triggers happy emotions led CRE8 DESIGN to translate the joyful company image into a young and lively pattern on a box that echoes with the product's main function – music streaming. The optimised design of the double cutout windows not only allows the speaker to be viewed from different angles, but it also gives the product a floating illusion which emphasises the light weight and portability of the bluetooth speaker

Design Agency: CRE8 DESIGN. Client: The Joy Factory. Nationality: USA.

Sound Box Package Design

···

Package design for a portable mp3 speaker. The design showcases the new-to-market product from 180 degrees.

Design Agency: newGStudio. Designer: Vladimir. Client: DEOS Style. Nationality: USA.

BASF Promotion Gift Package

BASF is a world's leading chemical company that believes in communication and interaction among people. This promotion gift is a speaker and the message is to amplify the voice of innovation. The design objective is to describe the relationship between chemical and human. The heart is composed of chemical compound and the centre says "Speak up and make surprises happen." Along the bottom of the illustration there is a series of squiggly lines joined by two people on either side. Not only do the wavy lines suggest the rhythm of a beating heart but also it shows the on-going communication of two people and explains BASF's motto: Bring the People and Chemistry Together.

Design Agency: one ZEBRA Limited. Creative Director: Clara Yeung. Designer: Clara Yeung. Client: BASF. Nationality: Hong Kong, China.

Philips Earphone Package

When the designer bought an earphone, he was intrigued with the amount of material used in the packaging and limited opening (plastic welded), so he thought an economical way, with good exposure and easy opening, and also highlights the main advantage of the product, the earhook.

Design Agency: pedrovidal. Designer: Pedro Vidal. Client: Philips. Nationality: Brazil.

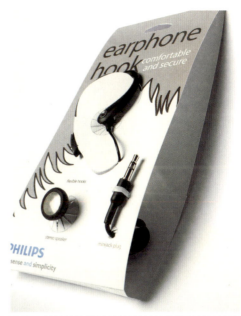

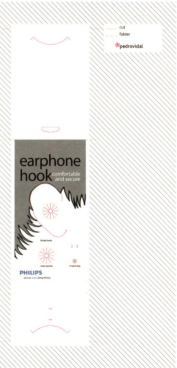

PULSE Headset Package Design

In order to create a pleasurable experience of opening a gift, CRE8 came up with these uniquely designed box sleeves. Leaving one side transparent for showcasing the product, the front and the side surfaces of the box are decorated with die-cuts and spot varnish techniques to deliver a refined texture.

Design Agency: CRE8 DESIGN. Client: Antec. Nationality: USA.

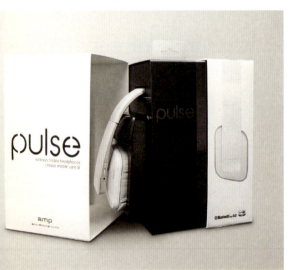

Icon7® VOX Touch Headset Package Design

Aiming to create a distinctive package that makes this award-winning product stand out in the high-end headset market, CRE8 DESIGN, with a limited budget in mind, adopted standardised technologies in an innovative way and transformed traditional materials such as paper and blister into a minimalist box with an iconic diamond-cut shape to display the product and highlight its value. The clean angular window and bold usage of the colour orange draw all attention to the main attraction, the VOX Touch. The gloss-varnished concentric line details on the black surface bring out subtle hint of the product's wireless sound superiority. The premium feeling is enforced further by keeping all primary surfaces clean – all the printed information appears only on the back of the box.

Design Agency: CRE8 DESIGN. Client: Icon7, Taiwan. Nationality: USA.

The Bottled Walkman

SONY

The Bottled Walkman

With the Sony waterproof MP3 player, swimmers can now enjoy the performance enhancing benefit of music while they swim. It's something every serious swimmer would want, but the challenge was getting the attention of this niche market. So FCB Auckland hijacked something they'll find in every gym over the world, vending machines, by combining the product with bottled water. In doing so they created a world first The Bottled Walkman. This simple packaging innovation gave a unique way to display the product, which instantly demonstrated its benefit.

Design Agency: FCB Auckland. Creative Director: James Mok. Designer: Melina Fiolitakis, Michael Braid, Kevin Walker. Photographer: Lewis Mulatero. Client: Sony. Nationality: New Zealand.

DEOS Earphones Package

Package design for DEOS Earphones. The cleverly designed instruction icons and neatly arranged product information keep the exterior clean and sleek, while the inner package provides a strong hold for the headset.

Design Agency: newGStudio. Designer: Vladimir. Client: DEOS Style. Nationality: USA.

Muzzio Cutedocs

Packaging design for earphones prepared in four versions. Each version has been prepared for a different type of earphones.

Design Agency: RiotHueLab. Designer: Lukasz Szejbut. Client: Vedia. Nationality: Poland.

Vedia SV500

Packaging design for earphones. Embossing of selected graphic elements of the design as well as golden Pantone have been used.

Design Agency: RiotHueLab. Designer: Lukasz Szejbut. Client: Vedia. Nationality: Poland.

CEF Earphone Package

When you buy the earphones, beautiful packaging is the first impression of your earphones. So the designer chose the beautiful geometry to show the cool of the earphone.

Design Agency: Dawn. Designer: Barton Williams. Nationality: UK.

"divide sound" Packaging

Package design for "divide sound", an audio device system which can divide out the sounds of instrument track from the music, so you can easily listen to the certain instrument sound you want to listen. This product is perfect for musicians who want to listen to the exact detailed sound of a certain instrument from the music. Basically, "divide sound" automatically recognises and divides out the individual sounds of instrument track from the music. Dividing concept is applied on the logo and the packaging for the earphone and the headset.

Designer: Hyun Dong (Ryan) Park. Nationality: South Korea.

EE Headset Package

The strange appearance of the package is the packaging design characteristics. Simple geometry type as a background, plus the text description of the headset. The geometry of the background and the text cleverly arranged in a strange box reflect the brand's taste.

Designer: Shayne Ward. Client: EE. Nationality: Norway .

Sweet Headset Package

Sweet is a specialised headset. The 2011 concept is aimed at young people. Considering youthful, dynamic characteristics of young people, the designer chose several candy colours to meet the annual philosophy of "Sweet". With irregular symmetry, this packaging is more youthful and lively, popular among young people.

Design Agency: Fancy. Designer: Jesper Green. Nationality: Denmark.

Swirl Headphone Packaging

The Swirl is designed with an emphasis on sturdiness and lasting durability. A robust, industrial in-ear housing made to withstand heavy everyday use. Signs of "Play" and "Pause" are used as graphic elements on the package, creating a simple yet unique design.

Design Agency: AIAIAI. Designer: Christian Zander, Kasper Nørlund. Client: AIAIAI. Nationality: Denmark.

Tracks Headphone Packaging

AIAIAI is a brand that highlights simplicity and minimalism. Designs for both its products and packages are simple yet unique. Tracks Headphone packaging is no exception; no more than three colours are used, yet a sense of liveliness is achieved, echoing the simple design of this product.

Design Agency: AIAIAI. Designer: Christian Zander, Kasper Nørlund. Client: AIAIAI. Nationality: Denmark.

Pipe Earphone Packaging

The design brief is to create a collection of packaging that is eye-cathing, displays the earphones and works together across the whole line of products.

Design Agency: AIAIAI. Designer: Christian Zander, Kasper Nørlund. Client: AIAIAI. Nationality: Denmark.

Pipe Earphone
Black

Pipe Earphone w/mic
Grey Gradient

Pipe Earphone
Grey Gradient

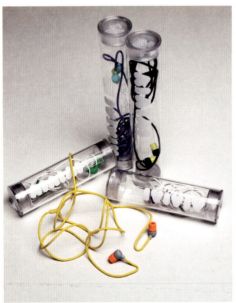

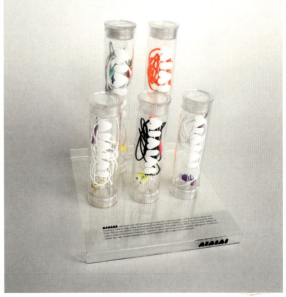

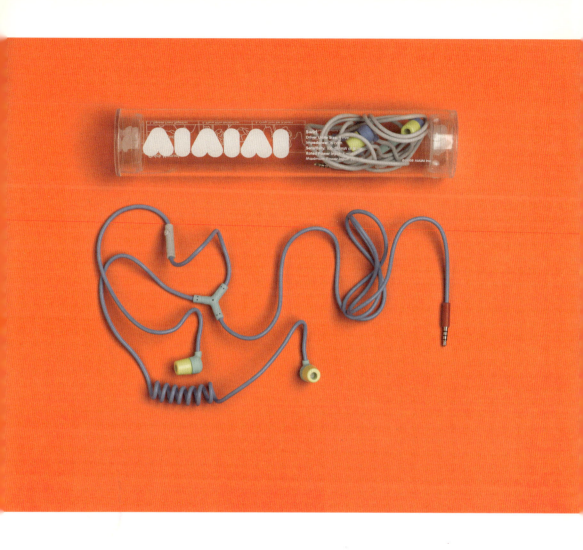

Tubes Earphone Packaging

The Tubes offers impressive sound quality and overall strength in an organic and ergonomic design. Made from lightweight and highly resilient materials this earphone displays an elegant coherency between the overall design and the individual parts.

Design Agency: AIAIAI. Designer: Kasper Nørlund, Peter Mix Willer. Client: AIAIAI. Nationality: Denmark.

Concert for one

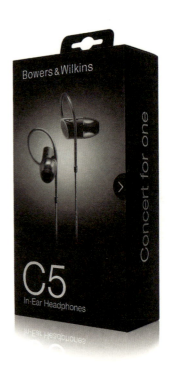

Bowers & Wilkins C5 Headphones and Accessories Packaging

After a successful project designing the accessory packaging for high-end speaker specialists Bowers & Wilkins, Burgopak were tasked with producing another packaging solution to hold their new noise-isolating C5 In-Ear Headphones. It was important for Bowers & Wilkins to retain their heritage as a premium brand as well as their reputation for producing high-quality products when moving into the in-ear category. This of course also had to be reflected in the packaging design – balancing chic aesthetics with intelligent structural design to securely hold the product in an eye-catching and retail-friendly box.

Design Agency: Burgopak. Creative Director: Dane Whitehurst. Designer: Hamish Thain. Client: Bowers & Wilkins. Nationality: UK.

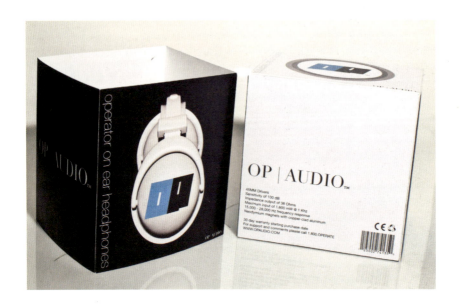

OP | AUDIO Headphones Packaging

....

The OP logo is an italic version of the typeface "OPERATOR". The typeface is formally and functionally minimalistic; two rectangular sticks of same width and height and three angles of turns.

Designer: Hyun Kyu Seo. Client: Fujifilm USA. Nationality: South Korea.

Psyko Audio Labs Identity and Packaging

Elevate Graphic Design worked with Psyko Audio Labs to develop their brand identity and create packaging for their 3D gaming headset line. Product graphics and word marks for the Carbon and Krypton models were also designed. By creating a distinctive graphic design solution, Elevate gave Psyko Audio the confidence to compete in a competitive market.

Design Agency: Elevate Graphic Design. Designer: Corey Brennan. Photographer: Stefanie Villeneuve. Client: Psyko Audio Labs. Nationality: Canada.

Idiosyncrasy Tune Emporium

Idiosyncrasy Tune Emporium's concept is derived from the influence music has on our personality and characteristics. The use of illustrative characters and bright colours are interpretations of the music genres they represent.

Designer: Steve Neuberger. Nationality: USA.

New York Funky

Different colours for different people! This packaging wants to show the headphones and care the product at the same time. A cut is made on the box to output the item but protected with PVC. The box has a handhole and base to choose the way to present it in shops.

Design Agency: jesus-lopez.com. Designer: Jesús López. Client: Noru. Nationality: Spain.

Crazy Zoo

Imaging a full colour box in a shop, but what is this? It is a surprise!
People would want to get the box and love it! Four different creatures
are created with different colours and personalities for each earphone;
but first, you need to get the box, because the product is not shown!

Design Agency: jesus-lopez.com. Designer: Jesús López. Client: Noru.
Nationality: Spain.

earBudeez Earbuds Packaging

Audiovox Accessories Corp gave JDA, Inc. an opportunity to reinvent the ME (Mobility Expanded) brand to appeal to a younger audience. Key drivers in earbuds purchased in the $10-$30 range were colour, style, and packaging – not performance. Drawing inspiration from sport shops, grocery stores, fashion boutiques, skateboard companies, and edgy graphics on energy drink cans. earBudeez – earbuds with big personalities using the earbuds themselves as the eyes of individual characters – was one of those ideas the designers thought "cute" but would never fly. Audiovox did go for it! Meet Bodie, Emo, Jay D., Jill, Skull Rojo and Zoie Jane.

Design Agency: JDA, Inc. Creative Director: David Jensen. Designer: Jerome Calleja, Dean Kojima, Stephanie Han. Client: Audiovox Accessories Corporation. Nationality: USA.

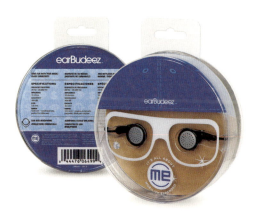

Pioneer Earphone Packaging Series

Pioneer took a fashionable and high-tech approach with their packaging and product colours. JDA had a unique challenge with this project. Instead of designing the structure of the packaging and creating a consistency for the brand, the requirement was to take the packaging designed in Japan, and make it relevant to the U.S. market. One of the challenges was to find a consistent spot for the Pioneer logo, which they determined to put on the bottom of the front of the package or label. They also took background graphics on the EQ, Steel Wheels and Loop series that didn't suit the brand in the U.S. market and gave them more shelf presence.

Design Agency: JDA, Inc. Creative Director: David Jensen. Designer: Dean Kojima, Maria Bautista. Client: Pioneer Electronics USA. Nationality: USA.

Klipsch/Mode M40 Headphones™

The packaging is meant to reflect the compelling relationship between fashion and function that is at the heart of this product. The background pattern was created from duplicating a wood grain. This was to allude to Klipsch's legendary wood-crafted, hi-fidelity speakers and make the connection between the company's headphone business and its history.

Design Agency: Klipsch Group, Inc. Designer: Matthew Miller. Photographer: Mark and Jen Halski. Client: Klipsch Group, Inc. Nationality: USA.

O2 Headphones Packaging

....

This product packaging project was created on the basis of helping creating a product, either fake or real, and being able to package this product in a way that was intriguing and made sense for the users. This project is a fake headphone brand called O2 headphones designed to help you retain twice the amount of sleep you actually get. The packaging is made from wood and is intriguing enough so the user would be able to keep it out on their bedside table to allow for no damage to be done to the headphones themselves. The package is in a shape of an earbud folded up and oxygen model when it is laid flat.

Designer: Kelsey Allen. Nationality: USA.

LG Bluetooth®

The extensive experience and understanding of the consumer electronics market helps Burgopak USA create packaging designs that consider brand, consumer and retailer. Packaging that differentiates a product from the competition, ensures its protection and level of security while remaining user-friendly and brand-conscious. Working with a broad range of materials and design processes suitable for the adequate protection of electronic devices, Burgopak USA create unique and bespoke designs, tailoring their patented sliding mechanism for feature-rich solutions and dynamic shelf presence, such as their designs for LG.

Design Agency: Burgopak USA. Creative Director: Hamish Thain. Designer: Mike Seiders. Client: LG Electronics Mobile. Nationality: USA.

Pixus Media Players Packaging

As a rule, names of gadgets are senseless and hard to memorise. In this project the designer tried to simplify recognition of a brand. By using numbers instead of complicated names, he made Pixus product line-up obvious and crystal clear. Alongside with the logo a unique font is made to match it.

Design Agency: Arriba! Media Group. Designer: Alexander Kryvonosov. Client: Pixus. Nationality: Ukraine.

Muzzio Grand/Muzzio MonsterHD

Packaging design for an audio/video player. Selective UV paint and embossing of selected graphic elements of the design have been used.

Design Agency: RiotHueLab. Designer: Lukasz Szejbut. Client: Vedia. Nationality: Poland.

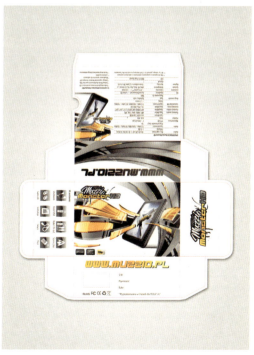

SiriusXM Packaging Redesign

Out of several current SiriusXM Satellite Radio agencies, JDA was awarded the project to update and create a new packaging system for SiriusXM radio products and accessories. One of primary goals was to make it easier for customers to find the right product and accessory to compliment and enhance their SiriusXM subscription. SiriusXM also wanted to tie the design to their website, which is clean with large fields of white, to the new packaging design.

Design Agency: JDA, Inc. Designer: Stephanie Han. Client: SiriusXM Radio, Inc. Nationality: USA.

Agatha

Packaging design for an MP4 player. Clean colour reflects the appearance of this product. The graphic design on the package reflects the MP4 technology. The key geometric element gives an intuitive message of the packaging: this is a player.

Design Agency: wonder. Designer: James Blunt. Client: Agatha. Nationality: New Zealand.

iPod in the Popsicle

The project is to create a simple promotion to be used throughout Brazil, for Kibon's Frutare popsicle line, during the most important sales period of the year: summer. It is a replica of the popsicle; EVA craddle and resin mold are used to protect the iPod against the low temperature and moisture. 10,000 iPods were distributed throughout Brazil. Whoever finds one will receive the accessories at home plus a real popsicle.

Design Agency: Bullet. Art director: Adriano Cerullo. Client: Unilever. Nationality: Brazil.

NRG Packaging

The task was to make a series of packages targeted at a young and progressive audience. Arriba! wanted to build the image of the energetic and contemporary brand. So they decided to use vivid graffiti-style illustrations to fulfill this goal. Every illustration on a package represents the main feature and a benefit of a corresponding media player. In addition to the package design, they have built a unique font to match NRG logo and to represent the name of a media player.

Design Agency: Arriba! Media Group. Creative Director: Max Burtsev. Designer: Alexander Kryvonosov. Client: NRG. Nationality: Ukraine.

Nano MP4 Box

Basically the company asked for a packaging for their MP4 players. The designer tried to make something different to all the packaging that you can find in Argentina.

Designer: **Daniel Selser**. Client: Nano. Nationality: Argentina.

MP3 Sport Live and MP4 Cool Live

Mp3 and Mp4 Live Series are designed for moving and cool people. The designer wanted to transfer to the customer movement and party sound...

Design Agency: jesus-lopez.com. Designer: Jesús López. Client: Noru. Nationality: Spain.

Vedia eReader K1

Packaging design for an ebook reader. Selective UV paint and embossing of selected graphic elements of the design have been used.

Design Agency: RiotHueLab. Designer: Lukasz Szejbut. Client: Vedia. Nationality: Poland.

Oh! Haus

After studying and seeing many products and brands in supermarkets and hardware stores, the designers decided to develop a line pop, fun and colourful for Oh! Haus. Who said the plumbers and the world of home electronics could not be cool?

Design Agency: Orange bcn. Designer: Aïda Font, Jordi Ferrandiz. Client: Greutor – Oh! Haus. Nationality: Spain.

Zune Digital Media Player

Swerve were asked to create packaging to reflect the warm, engaging and personal experience and to provide a unique moment of interaction and resonance between the consumer and the brand. Swerve recognised the need to explore a range of packaging concepts that layered the reveal of the product and orchestrated the theatre of first-time use. Form, proportion, materials and details were designed to evolve as moments of pleasure, hinting at the deeply personal nature of the Zune experience. The opening experience unveiled the personality of the brand as much as the product itself. Swerve worked closely with the Zune team and the manufacturers to ensure that the packaging was executed to the highest standards. Over three generations of players and two lines of accessories, Swerve helped Zune develop and amplify the DNA of the brand through the packaging design.

Design Agency: **Swerve**. Client: **Microsoft**. Nationality: **USA**.

Sony EyePet Promotional Packaging

The aim for this project was to design and produce an engaging structural packaging solution to securely hold the EyePet game software, literature and hardware. The pack was aimed at pan-European video game, technology and children's media journalists. Resembling a house-shaped pet carrier, the package features an intriguing pop-up mechanism where the EyePet character jumps up as the carrier opens from a hinged "roof". The package expands entirely, presenting the contents – the game's software, literature and hardware – in an enjoyable and organised manner. Burgopak worked on the EyePet project in conjunction with graphic and interactive design agency GR/DD, who developed the pack's graphics and initial pop-up concept. In order for the idea to be realised, Burgopak's structural design team developed a simple linkage and friction lock that allows the internal surfaces to move through 90 degrees and lock to form a flat table where the contents are securely held.

Design Agency: Burgopak. Creative Director: Burgo Wharton. Designer: Neil Usher. Client: Sony. Nationality: UK.

Sony EyePet & Invizimals Promotional Packaging

Held closed by a magnetic strip, the pack neatly unfolds open to reveal a bright and enticing introduction into the world of Sony's two new games. The presentation reveals a PlayStation®Eye camera, leading the recipient on to explore the EyePet and Invizimals contents separately through an intuitive cross-folding panel. Burgopak liaised with the manufacturing partner to ensure the design's die-line was optimised to reduce waste and keep within budget while considering manufacturing parameters. The design features high-quality finishing with double embossing, double sided print, matt laminate finish and spot UV. GR/DD produced the design's vibrant artwork and coordinated the initial structural concepts.

Design Agency: Burgopak. Creative Director: Dane Whitehurst. Designer: Hadley Baker. Client: Sony. Nationality: UK.

Welcome Kit BMW Motorrad

A high market value product, like a BMW motorcycle, demands a souvenir of compatible value. So, the designers chose another design icon, the iPod, as a gift, conditioned in a package with the earphones in the form of the motorcycle. It intensified the purchase experience and inspired, due to its exclusive content, the unique sensation of driving a BMW.

Design Agency: Bullet. Creative Director: Adriano Cerullo, Thais Mazelli. Client: Unilever. Nationality: Brazil.

Bem-vindo às melhores trilhas da sua vida.

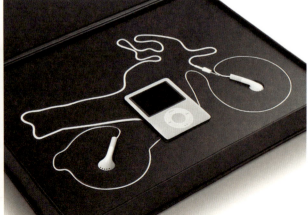

Pro Wrestling

This is a special project to create limited edition packaging for the restoration of the 1985 Nintendo hit, Pro Wrestling. The packaging includes a new three-dimensional logo combining elements of the classic title with the modern revamp. The bottom of the ring slides off to reveal the game cartridge inside.

Designer: Simon Blockley. Photographer: Jocelyn R.C. Client: Nintendo. Nationality: USA.

Dingoo Game Packaging

Branding and package redesign of Chinese video game Dingoo. The designer has been inspired by classic games package such as Sega Master System and Nintendo Entertainment System, and brand new consoles, like PlayStation Portable and Nintendo 3DS. In this package design, the look of modern games is merged with details of classic consoles.

Designer: Ricardo Ronda. Client: Dingoo Technology. Nationality: Brazil.

Retro Aviation Special Edition PSP

The main concept for the PSP is "Retro Aviation" and the idea for the packaging is "Vintage Suitcase". It was designed with a handle, a tag and vintage stickers. The remarkable thing about the packaging is that it has a various stickers' arrangement on every suitcase, which makes every single purchaser feel exceptional.

Designer: Fannie Cynthia Debby. Nationality: Indonesia.

XBOX 360 Gaming Products and Accessories

Swerve were asked to develop a new XBOX 360 package design to broaden their consumer base appealing to other adults and families as well as hardcore gamers with a packaging system that addressed the look and feel of a large range of XBOX products and accessories. Swerve recognised the need for the combination of messaging zones and physical forms to become instantly recognisable yet remain adaptable within a three-dimensional system. Swerve developed a simple set of forms that could be transposed, as the system grew, to products of different sizes and proportions. Subtle curves, consistent waisted silhouette, clean messaging surfaces, and simple arching colour panels used were able to convey a unified sense of friendly precision and consistency in the range of products. The overall look supported the extension to an empowering brand with broad appeal.

Design Agency: Swerve. Client: Microsoft. Nationality: USA.

Sifteo Cubes Packaging

Sifteo Cubes packaging is a playful and communicative showcase for the innovative new product. Lively graphics and colour pops beneath the Cubes grab the viewers' attention while securely holding them in place. A sleeve both protects and reveals the Cubes inside. The packaging is a cost-effective and eco-friendly solution.

Design Agency: NewDealDesign. Creative Director: Gadi Amit. Designer: Amy Yip, Maya Acosta, Barbara Stettler. Photographer: Mark Serr. Client: Sifteo Inc. Nationality: USA.

Digital
Watches

Nooka Paper Pack

....

Nooka is continuing its tradition of green packaging with the introduction of the Nooka Glue-less custom box. This paper packaging is truly eco-friendly, requiring minimal material, minimal labour and is biodegradable. The new packaging will be shipping with the Zub 40 line as part of Nooka's fall 2011 collection.

Design Agency: Nooka. Designer: Michael Ubbesen, Jon Patterson. Client: Nooka. Nationality: USA.

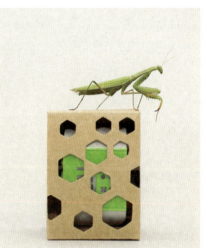

Nooka Reuseable Silicone Packaging

To continue with the tradition of innovation, Nooka proudly announced the debut of truly eco-friendly packaging with the SiliconeZone Nooka gem box. The new packaging is inspired by the Nooka gem shape, and is made from cooking grade silicone. Now proud Nooka owners can show off their watches, or warm up their favourite snack in the packaging.

Design Agency: Nooka. Creative Director: Matthew Waldman. Designer: Alexander Yoo, Paul Isabella. Photographer: Jon Patterson. Client: Nooka. Nationality: USA.

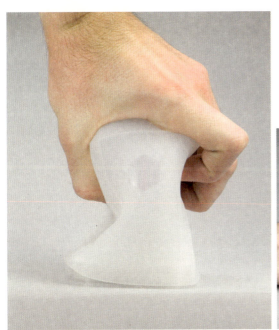

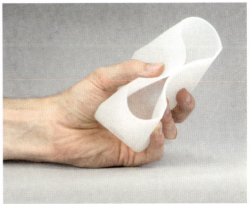

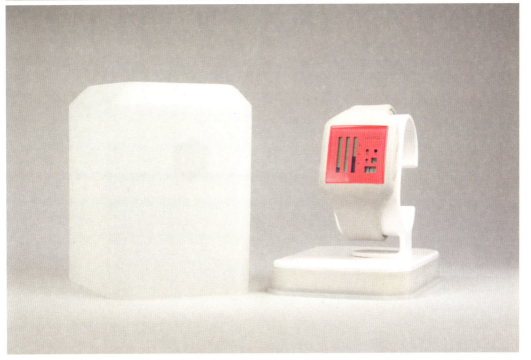

Nike Watch Series

This project constitutes a new proposal for Nike Watch Series, a sports watch model that was developed with the intention of being one of the new Nike products, regarding a rubber and plastic digital watch, easy to wrap around the wrist, with large and functional buttons, providing the user with a simple and direct usage of this product. Alongside the watch, a new packaging was developed, with the use of a trimmed cylinder, half transparent, for the user to see the watch and its colours, and half opaque, where Nike's logo and model description can be clearly viewed.

Designer: Tiago Russo. Client: Nike. Nationality: Portugal.

INDEX

Knoed Creative

Lexon Design

lg2boutique

Lifeworks Technology

Lomographic Society International

Make it Clear

María Laura Caballero Tejada

Mary Boyle, Connie Shim, Christina Xu

Matadog Design

McgarryBowen

Michal Marko

Multilase

NewDealDesign

newGStudio

Nicholas Menghini

Nick Reid Design

NO BODY Visual Communication Design

Nooka

one ZEBRA Limited

ONIS Design Studio

Orange bcn

Pantech

pedrovidal

Projectarthur

Quasidesigner.com

R2works

Rafał Czaniecki

rco design

Rene Kim

Ricardo Ronda

Rich Pageant

RiotHueLab

Rouge 24, Inc.

sanseliv

Shayne Ward

Simon Webbe

SONIC Design

Stas Bordukov

Steve Neuberger

Sublimio - Unique Design Formula

Swerve

Tatiana Rogatyuk

The Partners

Thinkdm2

Tiago Russo

Today Designers

Tomas Ashe, Claire Buckley

Twintip

wonder

Xiaoxi Li, Yun Wang, Emannuel Saka

Z3/Studio

© 2015 by Design Media Publishing Limited
This edition published in May 2015

Design Media Publishing Limited
20/F Manulife Tower
169 Electric Rd, North Point
Hong Kong
Tel: 00852-28672587
Fax: 00852-25050411
E-mail: suisusie@gmail.com
www.designmediahk.com

Editing: Kris Verstockt
Editorial Assistant: Liying Wang
Proofreading: Katy Lee
Design/Layout: Muzi Guan

ISBN 978-988-14123-1-7

Printed in China